SUPERNOVA 1987A: ASTRONOMY'S EXPLOSIVE ENIGMA

Written by
Russell M. Genet
Donald S. Hayes
Douglas S. Hall
David R. Genet

Published by the

FAIRBORN PRESS

Published in the U. S. A.
by the Fairborn Press
P. O. Box 7531
Mesa, AZ 85206

ISBN 0-944389-01-5

Library of Congress Catalog Card Number: 87-82773

Printed by Arcata Graphics Fairfield
Fairfield, Pennsylvania

Cover designed by Donald S. Hayes.

CONTENTS

PREFACE

When an astronomical event occurs which is as important as Supernova 1987A, the news media soon pick up the story and begin reporting the event to the public. Thus *Time* magazine featured Supernova 1987A on its cover, and has carried several major articles. Similarly, *Newsweek* and other weeklies have carried articles on Supernova 1987A. The major newspapers began reporting the story soon after the discovery, and have continued coverage since. The monthly science magazines have also carried the news.

The depth of coverage allowed by the constraints of the popular news media is limited. This means that the stories cover only the superficial events and a brief description of the astronomical importance of the phenomenon. Yet, there are many readers who would like to know far more about the event and its importance to astronomers. For them, a book—length discussion is appropriate. In the case of Halley's Comet, there was sufficient advance notice for books to be published by the time the comet became bright enough to be seen with the naked eye. In the case of Supernova 1987A, no such advance notice was possible.

About two months after the discovery of the supernova, we realized that there was a need for a book describing what it was and why it was important. It was clear, however, that for the book to be most useful, it had to be gotten out quickly. We took up the challenge, and you are holding the result in your hands. Our strategy was to combine background chapters with chapters based on interviews. The interviews were with experts active in the study of Supernova 1987A. The background chapters give the information needed to understand the nature and importance of supernovae, and were written on the basis of our experience teaching at the university level. We had to face the problem of writing about a phenomenon before all the answers were in; our response was to write a book which will prepare the reader to understand why the supernova is important, and to be able to follow the story as it unfolds. Further, we chose to emphasize the people doing the research and how they have reacted to the event. It is rare that the general public can see up close how scientists react to an exciting discovery, and this is one of the most important stories for this book.

This book is intended for a broad audience of those interested in the science of astronomy and how it is done. We assume only a little background on the part of the reader; nevertheless, to cover the nature of our current understanding of the supernova phenomenon requires a certain maturity of approach. We hope our readers find it to their liking.

The preparation of this book has benefited from the cooperation of a large number of participants in the drama and a number of experts in the study of supernovae. We want to acknowledge here our debt to these people; the book could not be what it is without them. They number too many to list here, but you may find them featured throughout the book in the text, in sidebars, and in pictures. Portions of the text were read critically by Dr. Sumner Starrfield and Dr. Stan Woosley; their comments have helped make it a more accurate description of the astrophysics of stars and supernovae.

Throughout the text are "sidebars" telling about many of these people and their work.   The choice of who to include had to be somewhat arbitrary, because of the breadth of study in this field today. We have attempted to include a spectrum of researchers.   Our sample should be considered to be representative at the most, and certainly not exhaustive.   To those many outstanding researchers who have not been included, we apologize, but we hope you are pleased with those we chose to represent you. Finally, the manuscript has been read by Dr. Raymond Bloomer, of the U. S. Air Force Academy, Dr. David Crawford, of Kitt Peak National Observatory, Julia L. Hayes, of the Avalon Corporation, Dr. Laurence A. Marshcall, of Gettysburg College, Dr. Gibson Reeves, of the University of Southern California, and Dr. Nathaniel White, of Lowell Observatory.   To them we give our highest appreciation for their comments and suggestions.

We thank the people who have contributed their interviews and suggestions; they have made it a better book.   But we take full responsibility for the contents; if errors remain, it is upon our shoulders which any blame must fall. Our hope, however, is that the book you are reading will accurately convey the great adventure of astronomy today!

CHAPTER 1

DISCOVERY!

The light from the supernova explosion in the Large Magellanic Cloud had been expanding for 170,000 years. It was actually a shell of light 170,000 light years in radius, yet only a few light years thick—an extremely thin shell. The leading edge of the shell had penetrated well into the large neighboring spiral galaxy, and now it neared the outer planets of a solar system—our solar system.

On February 23rd, 1987 the light from the exploding star in the Large Magellanic Cloud (LMC) had passed the outermost planets and the first photons on the leading edge were only hours away from intercepting the night side of Earth. The LMC was clearly visible in Chile, and at the Las Campanas Observatory, Ian Shelton loaded a 0.2×0.25 meter glass photographic plate into a 0.25—meter telescope, pointed it in the direction of the LMC, and opened the shutter. It was too soon; the photons from the supernovae had not arrived yet, but they were inside the orbit of Jupiter.

Ian was the Resident Observer at the University of Toronto Southern Observatory (UTSO). The University of Toronto had an agreement with the Las Campanas Observatory for use of their site. While Ian's job was to run the UTSO 0.6—meter reflector, he had asked Bill Kunkel, the Resident Scientist for the Las Campanas Observatory, if he could use their astrograph. The astrograph belonged to the Mt. Wilson and Las Campanas Observatories (funded by the Carnegie Institution of Washington). Bill agreed to let Ian use the astrograph, and this was the second night Ian had used it to take a photograph of the LMC. The first night's photograph had not turned out well, but Ian had things set up properly this time.

About an hour later, at 2:25 UT (Universal Time), Ian closed the shutter, removed the plate, and developed it. A much better photograph than the previous night. Ian looked the plate over carefully and didn't see anything unusual.

The first weak advance guard of the photons from the supernova had finally arrived at the third planet. The Large Magellanic Cloud was now overhead in New Zealand and at 9:26 UT the keen—eyed amateur astronomer, Albert Jones, took a careful look at it. For a year now, Jones had been observing three variable stars in the LMC. Each night, when it was clear and he had a chance to observe, he compared the brightness of each variable star with nearby nonvariable "comparison" stars. Jones observed and recorded the brightness variations of these stars over time. In doing this, he was helping professional astronomers understand the S Doradus—type variable stars. These stars are thought to be very young stars recently condensed from gas and dust, and their brightness variations provide important clues to their nature. The visual brightness estimates made by Jones, a very experienced observer, were quite accurate—within about 0.1 magnitudes (about ten percent). Jones did not see anything unusual, although his keen, highly trained eyes would have instantly spotted the supernova had it been very bright.

At 10:37 UT in Australia, Robert H. McNaught took a 1 minute and 58 second exposure of the LMC with a small camera mounted on a tripod. Rob and his friend, Gordon Garradd, normally took pictures of both the Large and Small Magellanic Clouds every clear night. They had been doing this since September 1986 when they began a photographic patrol for novae. Already they had found three novae and several variable stars of other types. They were the only people taking photographs of the Magellanic Clouds nightly as they searched for "new" stars—thus chances were good that they would be the ones to discover the supernova in the LMC. However, on this night, Gordon was "clouded out." Although Rob had to cut his exposure short, he did get a photograph—the first photograph or any other detection of the Supernova from Earth. This photograph was to be very important to astronomers, as it pinpointed how bright the supernova was at that instant in time.

Fate, however, kept Rob from being the discoverer of 1987A, although he literally had the discovery in his hands. His own words describe it best:

"I have often advocated giving cursory scans if time is not available for a complete search. Somehow, on that day I never looked at my photos. It was a long day, after a terrible nightmare (I hadn't had a nightmare for several years) which awoke me after a couple of hours. I got up and started several projects which I'd meant to do for some time. The first was to start keying in the coordinates of all the LMC variables to allow a computer search whenever Gordon phoned, instead of laboriously checking catalogues and calculating phases. Several times that day I thought I should check my photos, but I never did, and the episode is quite mysterious to me. Normally I search without a thought!"

Later that same night in New Zealand, amateur astronomers Melvin Thomas and S. Ryder were testing a new guider on the telescope at the

Beverly Begg Observatory. Located in Dunedin, New Zealand, near the south end of the South Island, the Beverly Begg Observatory is operated by New Zealand amateur astronomers. They decided to take a photograph of the LMC—a good way to test the new guider. They didn't develop their picture that night, so they didn't see the darkened spot on the negative.

The earth turned some more, and it was now midnight (0:00 UT) at Greenwich, England. A new day, February 24, 1987, was beginning.

In Chile, Ian Shelton loaded a new photographic plate in the 10—inch astrograph at Las Campanas, pointed it at the LMC, and opened the shutter. It was exactly 1:20 UT on February 24th, 1987. Immediately the photons from the supernova began activating the silver halide crystals on the plate.

During Ian's three hour exposure, the Night Assistant at the Las Campanas Observatory's 1—meter Swope telescope, Oscar Duhalde, stepped outside to look at the sky. Oscar noticed the new bright star near the 30 Doradus nebula.

Shelton finished the timed exposure at 4:20 UT and removed the plate from the astrograph. He then inserted a second plate and exposed it on a second field of stars. Ian, being a methodical observer, decided he should develop the plate before going to bed.

At 7:15 UT, Shelton pulled the LMC plate out of the wash and looked at it, immediately noticing a spot where there should not have been one. Could it be a defect in the plate? Shelton puzzled over possible explanations before he stepped outside and looked up at the Large Magellanic Cloud. There, not far from 30 Doradus, the Tarantula Nebula, was a bright star which Shelton knew did not belong there. Now the photon message from the exploding star had been comprehended by a human being.

Ian was anxious to share his discovery, so he walked over to the nearby 1—meter Swope telescope where astronomers Barry Madore and Robert Jedrzejewski as well as Oscar Duhalde were on duty. What, asked Ian, would a 5th magnitude object in the LMC be if it was there tonight but not last night? Barry Madore said, as Ian had expected, that it had to be a supernova! Oscar Duhalde pointed out that he also had seen something. Ironically, he had been so busy that he hadn't had a chance to tell the astronomers.

Ian Shelton, Barry Madore, Robert Jedrzejewski, and Oscar Duhalde all trooped out of the Swope reflector's dome and stood outside on the bare Chilean mountaintop looking up at the LMC. Now the light from the supernova fell on four pairs of comprehending eyes! Barry immediately began calling Brian Marsden's[*] numbers (there are four of them). One number had a tape recorder on it, but it took eight rings to trigger it, and the international phone system would not let it ring that many times.

---

[*] Brian Marsden was the international coordinator for reporting astronomical events an urgent nature (see page 6).

Figure 1 (Previous page).   Two photographs taken with a small camera mounted on a tripod.   The top, pre—discovery picture was taken on February 22, 1987, at 11:07 UT.   The bottom picture was taken on February 23, 1987, at 10:37 UT.   The supernova on the bottom picture (pointed to by the arrows) is easily visible.   The bottom photograph is the record of the first detection of the light from the supernova.   (Photographs by R. McNaught.)

It was clearing in Nelson, New Zealand, and at about 9:00 UT on February 24th, 1987, Albert Jones swung his 0.3—meter telescope for a look at the three variable stars he was studying in the LMC.   There in the same viewing field was a very bright blue star that did not belong there!   Jones got his star charts out, and noted the position of the new star relative to other stars.   But before he could determine its actual brightness by comparing it with other nearby stars of known brightness, clouds moved in.

Jones waited awhile for the clouds to clear, and when they didn't, he called his old friend and fellow amateur astronomer, Frank Bateson, and gave him the position of the new star and a conservative (faint) estimate of its brightness.   Frank called up some other observers, including amateur astronomers further north in New Zealand, and astronomers at the Anglo Australian Telescope (AAT) in Australia.   The AAT astronomers called Tom Cragg, who called Rob McNaught.

After he got off the phone with Frank and went back out to the telescope, Jones saw the clouds were breaking, and soon the skies cleared. On February 24th, at 10:10 UT, Albert Jones' observations established that the visual magnitude of the supernova was 5.1.   He called his friend Frank Bateson again with this information.

Not knowing if other alerted observers had clear skies, Jones continued observing the supernova another 4 hours, by which time the brightening seemed to have leveled off somewhat.   He took an hour break for a short nap, and then observed the supernova until dawn.   He finished off with a look at Comet Wilson 1986l.   It had been quite a night!

When Rob McNaught received the word from Frank Bateson, via intermediaries, he swung into action.   Pre—discovery photos showed a 12th magnitude star at the position of the supernova.   A check of the photograph taken the previous night showed that the supernova had already started its rise—there it was at magnitude 6.1—a heart wrenching moment for Rob!   Rob, however, was much too busy to worry about what might have been, and he used the Aston Hewitt Schmidt Satellite Camera and the measuring engine at the AAT to accurately tie down the position of the supernova.   It appeared to be coincident with the position of a 12th magnitude B3 type supergiant, Sanduleak −69° 202 (the designation in Nicholas Sanduleak's star catalog).

THE WORD GOES OUT

Priority of discovery is important in science. Lifetimes are devoted to advancing knowledge, and when a breakthrough occurs, it is vital that proper credit be given. As a number of scientists are often working in the same area, independent discoveries of the same thing are not unusual. Science has a strict rule about discoveries; they must be made public to receive credit, and generally, the first person to make the discovery public receives the credit for it. If someone were to make a discovery and not reveal it, hoping to polish it up a bit or go on to further discoveries without alerting the competition, and someone else discovered it independently and made it public first—especially if they got it in print first in a scientific journal—then they would be given credit for the discovery. If this were not the policy in science, then results would not be reported promptly and progress would suffer.

Discoveries in astronomy are, in the main, treated like other discoveries in science. The critical action upon figuring out, for example, some new way a star behaves is to submit a paper on the subject to a recognized journal. The date the paper is received by the journal editor is considered the date it is made public, and it will have priority if the same idea has not been received earlier.

There is, however, a special difficulty in the case of sudden discoveries made while observing, especially when it is vital that these discoveries be communicated quickly to other astronomers so they can also observe what may be a transient object with other telescopes and different instruments. If, on the one hand, the discovering astronomer were just to start calling other astronomers, in the ensuing rush to observe it might not be apparent who made the first call and when. On the other hand, if someone doesn't make the calls, astronomers would not be alerted and important observations would be missed. There is a single someone to make these calls, and he is Brian Marsden.

Brian Marsden runs the Central Bureau for Astronomical Telegrams of the International Astronomical Union (IAU). He does this from his office in Cambridge, Massachusetts, where he works for the Harvard–Smithsonian Center for Astrophysics (CfA). The basic idea behind the Central Bureau for Astronomical Telegrams is that astronomers alert Brian of any urgent discoveries, preferably via telegram, telex, or electronic mail, and Brian in turn alerts all the astronomers around the world that subscribe to the bureau's service. It is Brian's job to keep track of who calls when and to judiciously decide who should get credit for a new discovery. Most of the discoveries reported are of new comets, asteroids, novae, and faint, far–away supernovae.

The morning of February 24th seemed like any other morning as Brian drove to work and walked up to his office by the old Harvard College Observatory telescope. He barely had time to check the overnight electronic mail when the telex machine came alive and typed a fairly brief message. Brian glanced at the clock; it was 8:56 AM Eastern Standard

Time (13:56 UT). He tore the message off the telex and sat down to read it. This is exactly what it said:

ASTROGRAM CAM

MSG. 087
DT: FEBRUARY 24, 1987

TO   CENTRAL BUREAU FOR ASTRONOMICAL TELEGRAMS
      SMITHSONIAN ASTROPHYSICAL OBSERVATORY
      CAMBRIDGE, MA 02138

      IAN SHELTON OF THE UNIVERSITY OF TORONTO STATION IN
THE LAS CAMPANAS OBSERVATORY, CHILE REPORTS A POSSIBLE
SUPERNOVA IN THE LMC NEAR 30 DORADUS AT RA(1987) 05H
35.4M AND DEC -690 16M. APPARENT MAG. 4.5, OBSERVED AT
APPROXIMATELY 0800 UT.

(SIGNED) W. KUNKEL
LAS CAMPANAS OBSERVATORY
24 FEBRUARY 1987, 1340UT

When Brian finished reading it, he sat and thought for a couple of minutes. It seemed a rather non—commital message. The phone rang. It was Barry Madore calling from Chile. Barry said the supernova was definite.

While Brian was talking to Barry, the other phone rang; it was Rob McNaught calling from Australia. This call, answered by Brian's secretary, Donna Thompson, provided an estimate of the supernova's brightness. Had Rob made an independent discovery? This was not clear.

Meanwhile, sometime after 14:00 UT (9:00 AM EST) Ian Shelton had called Donald Fernie in Toronto. Don, the Director of the David Dunlap Observatory, passed the word on to Bob Garrison, the University of Toronto Southern Observatory (UTSO) Director. It didn't take Bob more than a couple of minutes to realize he had a big job on his hands. Bob gave the media a few calls, and soon there were three camera crews in his office.

When Brian's assistant, Dan Green, arrived in the office, he was given the task of preparing and transmitting the discovery telegram (which went out at 10:00 EST), while Brian himself started work on an IAU Circular. While doing this, Rob called again, and Brian found that Rob had heard of the supernova from Frank Bateson in New Zealand, but it was unclear whether Frank had heard from Chile, or if someone in New Zealand had made an independent discovery.

Suspecting the latter, and although it was only 4:30 A.M. in New Zealand, Brian felt it was necessary to place a call to Frank. So, at 10:30 AM Eastern Standard Time (15:30 UT), Brian was talking to Frank in New Zealand. Yes, it was a completely independent discovery by amateur astronomer Albert Jones. Brian knew Jones to be a highly competent and reliable observer; his independent discovery had allowed the first photoelectric measurements to be made by New Zealand amateur astronomers W.S.G. (Stan) Walker and Brian Marino at the Auckland Observatory, and had alerted the Australians while it was still dark there.

Figure 2. Brian Marsden communicates over the computer network. For the early days and weeks of the supernova, astronomers anxiously awaited circulars from Brian, who issued several of them every day. For the first time, there were even weekend circulars. As Stan Woosley said for all astronomers "Brian was a real hero."

Brian Warner, an astronomer from South Africa, was visiting at the University of Texas. When he heard about the supernova, he contacted the South Africa Astronomical Observatory in time for them to obtain the first spectra of the new object, late on February 24th.

Further telephone calls from Robert McNaught aided Brian's attempts to complete the first IAU Circular. These calls brought the first accurate position of the supernova, as well as the first identification of the likely progenitor—a twelfth magnitude star. Rob had also developed his films of the night before, found images of the supernova, and estimated its brightness. Finally, there was word of the first spectroscopic results from South Africa and the incorrect conclusion that the supernova was of

Type I. All this was placed in Circular No. 4316 of the Central Bureau for Astronomical Telegrams of the International Astronomical Union. Those astronomers who subscribed to the computerized version of the service received their circular within minutes, while the rest got it via air mail.

The Earth had been receiving photons from the supernova for one day. All around the southern hemisphere where the LMC could be seen, telescopes large and small were either observing the supernova or waiting until darkness came their way so they could. In orbit around the earth, the veteran space telescope called the International Ultraviolet Explorer (IUE) was beginning to observe the supernovae. The word was out.

IAN KEITH SHELTON

Born in Winnipeg, Manitoba, Canada in 1957, Ian Shelton had an early interest in astronomy. He built his own back—yard observatory where he spent many hours observing planets, stars, and galaxies. Ian graduated from Garden City Collegiate in Winnipeg, and attended the University of Manitoba, Winnipeg, where he studied Physics and graduated (B.Sc.) in 1979.

In 1981, Ian applied for a job as a Resident Observer at the University of Toronto Southern Observatory (UTSO). Robert F. Garrison was the UTSO Director, and his assessment of young Shelton was positive. Ian got the job and soon he found himself on a remote mountaintop in Chile. Ian really liked observing, and he was good at it. A methodical, meticulous, and responsible person, he was well suited to the demanding tasks placed on him. Unlike many astronomers who yearn for civilization after only a few weeks on an isolated mountaintop, Ian thrived in this unusual environment, and he stayed on the mountaintop month after month, even preferring to spend some of his vacation time on the mountain.

After two years in Chile, Ian thought it was time for something else, and he returned to Canada and studied for a while at the University of Waterloo, not far from Toronto, and then worked at the Museum of Man and Nature in Winnipeg, Manitoba. However, Ian had the astronomy bug, and he wanted to get back into observing, so in 1985 he applied for the UTSO Resident Observer job again. Bob Garrison was still the UTSO Director, and was delighted to get Ian again for the key job in Chile.

Ian did not bask in the glory of his discovery. He was the Resident Observer at an important southern observatory. Ian was determined to make as many observations of the supernova as possible. He was so busy he didn't have much time to think about his discovery.

Towards the end of April, Bob Garrison brought Ian back to Toronto for a celebration of the discovery. Bob had reserved the largest hall at the University of Toronto, and on April 24th Ian stood before a standing—room—only crowd of 1700 and gave a modest 20—minute talk. The standing ovation came not only from the University of Toronto's faculty and students, but from the citizens of the city of Toronto itself. They were proud of the extension of Toronto in Chile where the message of the exploding star had first been understood by one of their own fellow Canadians.

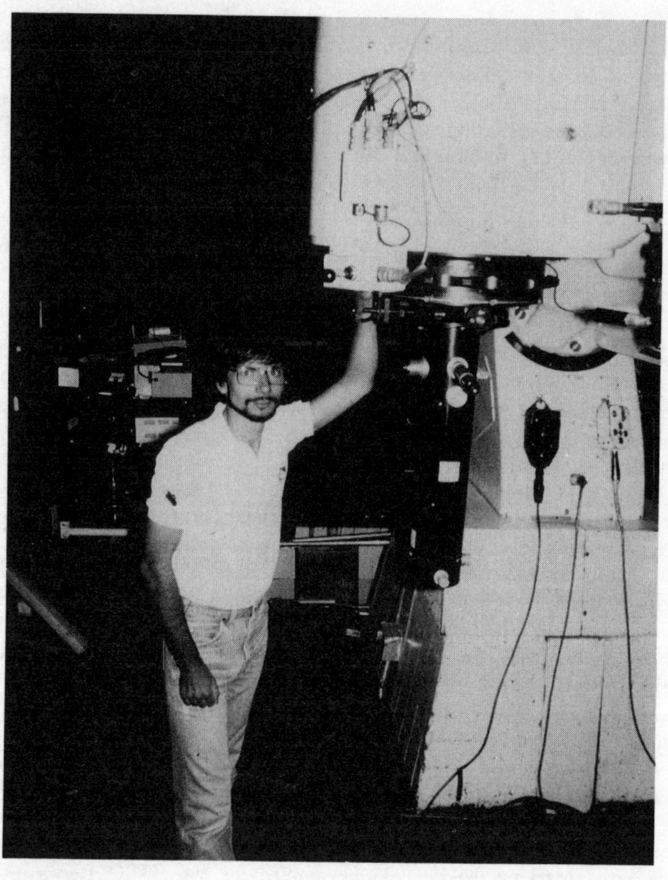

Ian Shelton and the 0.6—meter telescope of the University of Toronto Southern Observatory (UTSO). Ian has spent four of his last 6 years taking care of this telescope, making observations with it, and helping visiting astronomers use it to make observations. The combination of Ian's expertise and hard work, the 0.6—meter telescope, and the superb astronomical conditions on this remote mountain top, have produced outstanding astronomical data for the University of Toronto.

UNIVERSITY OF TORONTO SOUTHERN OBSERVATORY (UTSO)

The University of Toronto Southern Observatory at Las Campanas, Chile. This picture (provided by Robert F. Garrison), is looking north from the 1—meter Swope telescope belonging to the Carnegie Institution. On the right is the UTSO 0.6—meter telescope. On the left is the UTSO Service Building. Further out, and in the center, is the small building that houses the 0.25—meter astrograph, belonging to the Carnegie Institution. On the top of the mountain and slightly to the left is the large, 2.5—meter du Pont telescope, which also belongs to the Carnegie Institution.

The University of Toronto operates the 0.6—meter telescope in the Atacama Desert in Chile. Although of modest size, this telescope is well instrumented. The instrumentation includes three photoelectric photometers and a spectrograph, designed by Robert F. Garrison, used to classify stellar spectra. Recently, a Charge Coupled Device (CCD) camera has been added. On—site microcomputers are available for data reduction, and a short—wave radio allows direct communications with Toronto.

In addition to the 0.6—meter telescope, which is housed in a conventional dome, UTSO has a service building where the Resident Observer and any guest observers stay. In addition, there is work space and a dark room in the service building.

UTSO is supported and directed from Toronto. While Chile is far away, the UTSO Director, Robert F. Garrison, and the Resident Observer, Ian Shelton, stay in contact via shortwave radio.

One advantage of having a small telescope is that instead of doling out observing time a night or two at a time, as seems to be the custom on large telescopes, long–term observational projects can be undertaken that require many nights of observing, or take short intervals of time spread over many months or even years. This long–term capability is particularly important in the study of variable stars, a specialty at the University of Toronto.

UTSO is located on the grounds of the larger Las Campanas Observatory, of the Carnegie Institution. The 0.25–meter astrograph used by Shelton to make the discovery of the supernova belongs to the Las Campanas Observatory.

ROBERT F. GARRISON

Robert F. Garrison, Professor of Astronomy at the University of Toronto, and Director of the University of Toronto Southern Observatory (UTSO) (Photo by Russell Genet).

Born in Aurora, Illinois, in 1936, Robert F. Garrison had an early interest in mathematics and science. As an undergraduate Bob attended Earlham, a small Quaker college in Richmond, Indiana. While much too small to have an astronomy department, one of the physics professors, Clifford Crump, was a retired astronomer, and he was instrumental in stimulating Bob's interest in astronomy. Crump was the prime mover in getting Professor Perkins to donate funds for the Ohio Wesleyan Perkins telescope—now the 1.8—meter telescope at the Lowell Observatory. Crump was the first director of Perkins Observatory in 1922.

Bob continued his education at the University of Chicago, where his thesis adviser was W. W. Morgan, a leading spectroscopist and international expert on the spectral classification of stars. Bob became an expert himself in stellar spectroscopy, spectral classification, and galactic structure. On receiving his PhD in 1966, Bob went to the Mt. Wilson and Palomar Observatories for two years as a post—doctoral fellow where he worked with Armin Deutsch and Philip Keenan on the spectra of Mira variable stars, and then he went to the University of Toronto.

At the University of Toronto, which has a large astronomy department, Bob started as an Assistant Professor, and has since advanced to the rank of full Professor, and is Director of the University of Toronto Southern Observatory (UTSO). Bob was involved with UTSO from its inception, initially sharing the work with Rene Racine, now at the University of Montreal. Racine took care of setting up the telescope and designing and building the first photometer. Bob took care of the logistics of the Observatory and the design of the spectrograph. When Racine left, Garrison took over general direction. Eventually, Bob was able to have a Resident Observer to make it easier to run UTSO from Toronto.

THE UNIVERSITY OF TORONTO AND THE DAVID DUNLAP OBSERVATORY

Clarence Augustus Chant, a professor at the University of Toronto, talked the University into establishing an Astronomy Department in 1904, with himself as the only professor. Building a telescope for the department proved to be somewhat more difficult, but thanks to the generosity of the widow of David Dunlap, a 1.8—meter telescope was built for the University and dedicated as the David Dunlap Observatory (DDO). At the time it began operation (1935) it was the world's second largest telescope (after the Mt. Wilson 2.5—meter), and today it is still Canada's largest.

With this magnificent telescope in hand, the Department of Astronomy prospered, and frontier research in stellar spectroscopy, variable stars, and other areas was accomplished. Besides large—telescope astronomy, the University of Toronto has had a tradition of making contributions with small telescopes, particularly in the area of variable stars. The many accomplishments with their 0.6—meter telescope at the University of Toronto Southern Observatory (UTSO) is entirely within the tradition of the department.

The David Dunlap Observatory (DDO) of the University of Toronto. Inside the dome on the left is the 1.8-meter telescope, while on the right is the main building with several domes for smaller telescopes. The DDO's Director is J. Donald Fernie (Photo credit: the University of Toronto).

LAS CAMPANAS OBSERVATORY

The Las Campanas Observatory is the southern and now main station of the Carnegie Institution of Washington's Mt. Wilson and Las Campanas Observatories. With headquarters on Santa Barbara Street in Pasadena, California, the Mt. Wilson and Las Campanas Observatories have long been in the forefront of astronomy. George Ellery Hale built the first modern large reflectors on Mt. Wilson. The 2.5-meter Hooker telescope was used by astronomer Edwin Hubble to discover the expansion of the universe.

As the city of Los Angeles grew, the skies at of Mt. Wilson became increasingly bright, and the Carnegie Institution of Washington began to look for a dark site that would allow them to continue work on faint galaxies. A search of many sites around the world led them to a desert mountain site in northern Chile. Clouds are rare, rain is even rarer, and the atmosphere is clear and steady. While the area is rather remote with vegetation, lakes, and rivers being noteworthy by their absence, it is an astronomer's paradise, perhaps the best location on the planet. It is not surprising that nearby are two other major observatories, the Cerro Tololo Interamerican Observatory (CTIO), and the European Southern Observatory (ESO).

Shortly after establishing the site, and with the help of the Chilean government, construction began on the first telescope, the 1-meter Swope reflector.   Named after Miss Henrietta H. Swope, a member of the Mt. Wilson scientific staff who donated money for its construction, the telescope began operation in 1971.   At about the same time, an agreement was reached with the University of Toronto that allowed them to install and operate a 0.6—meter telescope at the site.

In 1976, the 2.5—meter du Pont telescope began operation.   Made possible by a gift from Mr. and Mrs. Crawford H. Greenewalt, Mrs. Greenewalt asked that the telescope be named after her father, Irenee du Pont.   This well—instrumented, highly modern telescope would not have the problems of light pollution that had affected Mt. Wilson; the nearest city (and not a large one at that) is over 100 miles away!

ALBERT JONES

Born in Christchurch, New Zealand, in 1920, Albert Jones moved with his family to Timaru in 1926.   Timaru is a small town on the east coast of the South Island—about 100 miles south of Christchurch.   Jones attended the Timaru Boys' High School from 1933 to 1936, graduating in 1936.   While there, a Geography Master who was knowledgeable in astronomy sparked a heavenly interest in young Albert.   He was too shy to approach the Master with the many questions he had, but after graduating from school, a chance meeting in the public library helped break the ice, and the Master helped Albert get started in astronomy.   One day, Albert happened to see a request in the local paper for reports on the Aurora Australis (the southern lights) to be sent to the Carter Observatory (later to become New Zealand's national observatory).

Young Jones, then only 19 years old, made a few observations of the Aurora Australis, and sent in his report to the Carter Observatory.   Soon he received an appreciative letter from Murray Geddes, then Director of the Observatory, and that encouragement so fired his enthusiasm that he became a regular observer of aurora.   Later, after he had acquired a small reflector, he added comets and variable stars to his program and joined the Variable Star Section of the New Zealand Astronomical Society.   This section was founded and headed by Frank Bateson.

Jones enjoyed observing variable stars.   He started with Mira variables (reddish—colored giant stars that make large changes in brightness over the course of about a year).   Later Albert added R CrB variable stars.   These highly unusual and rather rare stars are normally bright but sometimes they literally "soot up".   This affect is caused by carbon clouds forming in their atmospheres that block most of the visual light.   Jones reported these sudden plunges into darkness so that professional astronomers with their larger telescopes could study their strange behavior more closely.   Later Jones became "hooked" on dwarf novae—stars that are normally faint but, on rare occasions, suddenly brighten dramatically only to fade back to normal brightness after a few days. This is a field where amateurs help professional astronomers by monitoring the stars and alerting interested observatories when outbursts are detected.

One winter morning in August, 1946, Albert was observing variable stars when he noticed a comet near one of the variable starfields. He dutifully reported it.   As luck would have it, his report was the first.   It became Comet Jones 1946 VI.

Albert Jones stands beside his home—made telescope.   Albert has been observing stars for over 40 years and is one of the most accomplished variable star observers in the world.   Albert instantly spotted the supernova when he swung his telescope in the direction of the Large Magallenic Cloud.   (Photo provided by the Nelson Evening Mail.)

In 1948, Jones built his 0.3—meter, f/5 reflector, still in use almost 40 years later.   The tube is square, made with aluminum angles, braced, and covered with Pinex Softboard—a thick wallboard used in the construction of houses.   The fork—mounted telescope runs on small trolley wheels.   The telescope is stored in a shed, then wheeled out into the open for use.   The mounting has no drive or setting circles—objects are located in the finder first.   For variable stars and comets, a fairly low power with a wide field is used nearly all the time, so a drive would not be useful anyway.

As the years went by, Albert continued his job during the day as a breakfast food miller at the Timaru Milling Company. At night, he continued his observations of variable stars and comets with his homemade telescope. In 1964, he moved to Nelson, located at the north end of the South Island, where he was a partner in a grocery/delicatessen. Of course, the telescope was moved also, and the observations continued. Albert is now retired, and is observing more than ever. When he discovered SN 1987A, he was one of the most experienced variable star observers in the world, with some 350,000 observations to his credit—an incredible record!

While always a modest, even shy person who preferred quietly observing the stars, Jones could not escape the well—deserved recognition that was sent to him from around the world. A few of these awards include the Murray Geddes Memorial Prize of the New Zealand Astronomical Society (1945), the Donohue Comet Medal of the Astronomical Society of the Pacific (1946), Fellow of the Royal Astronomical Society (1947), the Mechaelis Gold Medal (1956), the Jackson—Gwilt Medal of the Royal Astronomical Society (1960), Fellow of the Royal Astronomical Society of New Zealand (1963), Member of the Variable Star Commission of the International Astronomical Union (1965), the Merlin silver Medal of the British Astronomical Association (1968), and the Astronomical Society of the Pacific's Comet Medal (1973).

Albert Jones thinks this is all an unnecessary fuss over a few hours spent quietly at the telescope. The Mayor of Nelson disagrees, awarding Albert the Nelson 1987 Achievers Award. Also, the Governor General of New Zealand informed Albert, just before the manuscript for this book was completed, that "The Queen has been graciously pleased on the occasion of Her Majesty's Birthday to confer on you the honor of Officer of the Civil Division of the Most Excellent Order of the British Empire."

Sir Albert Jones remains unaffected by all the fuss. He continues to observe variable stars with his home—made telescope.

ROBERT H. McNAUGHT

Born in Scotland in 1956, Rob has had a passion for the skies from the age of 7. His first introduction to astronomy is vividly remembered. On being presented with an unwanted book for good attendance at Sunday school, he swapped it for "Timothy's Book of Space" and has ever since wondered what he'd be doing if he hadn't!

Before the age of 10, many hours were spent at night kneeling on his bed staring out a south window watching the constellations wheel across the sky. A fireball seen passing close to Orion later led to a keen interest in meteor observing. Other early memories include the majestically bright "Echo" satellites watched on every possible occasion. Comet Bennet in 1970 left a lasting impression. Alas, this early romance had to bow to growing up.

Upon graduation in 1981, Rob took up a post at the University of Aston, in Birmingham, England, as a research assistant in the Earth Satellite Research

Unit, operating the 0.6—meter f/1, Hewitt Satellite Camera, (for Birmingham located at the Royal Greenwich Observatory) in the beautiful surroundings of Herstmonceux castle in southern England. Rob made prolonged use of the extensive astronomical library housed in the castle. In 1984, he was transferred to operate the sister Hewitt Satellite Camera at Siding Spring Observatory in New South Wales, Australia.

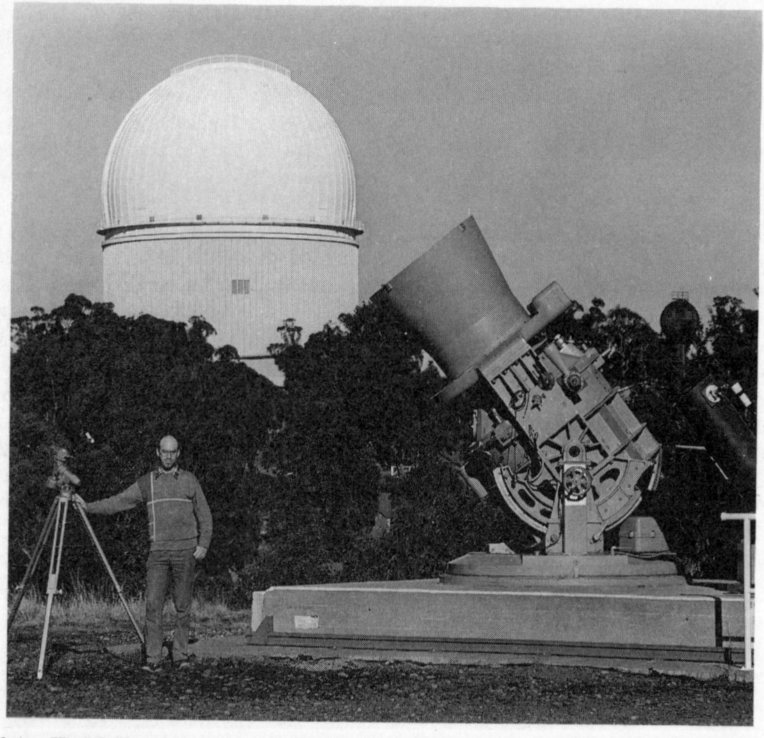

Robert H. McNaught stands beside the camera he used to photograph SN 1987A (The small black camera on the tripod near Rob's right hand). Behind Rob is the Anglo Australian Telescope (AAT), and to Rob's right is the Hewitt Schmidt Satellite Camera. On the far right is a 0.3—meter Dobsonian telescope. (Photo by R. McNaught.)

As a committed amateur astronomer, he has had marvelous good fortune and has not wasted the astronomical opportunities offered him. Since his teens, Rob has expanded his observational interests to include orbit determination, visual and photographic comet, nova and supernova search, and visual observation and historical research on cataclysmic variables.

CHAPTER 2

INTRODUCTION

The discovery of the brightest supernova in over 300 years, on February 23, 1987, upset the usually calm deliberations of the community of astronomers. For some of them, the events were revolutionary, and their lives changed considerably in the months that followed. The Supernova 1987A discovery is an example of that rare celestial event which excites the entire astronomy community; it is so important that the excitement even rubs off on the news media and the general public. The last such event was the passage of Halley's comet a year ago, and, for most astronomers, Halley's comet was not nearly as important as is Supernova 1987A. While comets are interesting, bright ones are not that infrequent. A bright supernova is truly rare. More significantly, in many ways supernovae are at the heart of astrophysics, which is itself at the heart of astronomy.

What is a supernova? It is the explosion of a star at the end of its lifetime. If we measure the intensity of an explosion by its total energy output, then only quasars and the "Big Bang" out—perform a supernova. But a quasar is a portion of an entire galaxy, not just a single star, so that comparison isn't really fair. The "Big Bang" was that biggest—of—all explosion which happened about 10 to 15 thousand—million years in the past, but which involved all of the mass in the entire universe, so here the comparison is even more unfair. One can appreciate how gargantuan a supernova explosion is by realizing that, during the month or so when it is at maximum brightness, the star can shine as brightly as the entire galaxy in which it resides. An extreme example is supernova SN 1937A becoming 100 times brighter than the galaxy itself (IC 4182).

As we will explain, the energy released in the form of visible light, x—rays, and radio waves is but a small fraction of the total energy released: ten times or maybe a hundred times more energy is released in an incredibly intense, brief shower of invisible subatomic particles called neutrinos.

Figure 1. *A galaxy with and without a supernova.*

How bright a supernova appears in the sky depends upon how far away it is.   If one occurred within a few thousand light years of the Earth, its apparent brightness would exceed that of any other star, and even exceed that of the brightest planets, Venus and Jupiter.   When this bright, it would become visible by day; this has happened a few times in recorded history.   Generally, a supernova will stay near its maximum brightness for a month or so, and then slowly fade away over a period of a couple of years.

Why is a bright supernova so important? It is important because supernovae are central to our understanding of how stars, galaxies, and even matter itself evolve, and a supernova must be bright to be within the range that allows us to study it in detail. Faint supernovae, mostly fainter than 10th magnitude, are commonly seen in other galaxies; they are faint because other galaxies are very far away. Such faint supernovae are helpful to our study of the phenomenon, but many of the crucial observations we need to make are impossible because of their faintness. The last bright supernova occurred just before the invention of the telescope, so astronomers haven't had the chance to apply modern technology to the study of bright supernovae before this year. Previously impossible observations became possible, and they are at the heart of astrophysics. This is why the astronomical community is so excited, why it has changed the normally fixed observing schedules on all the major southern telescopes, why astronomers have been seen more often on airlines flying to the southern hemisphere, and why some astronomers are shouting at each other.

Besides their importance to understanding the death throes of a massive star, supernovae are important to study because of the vital role they play in the creation and dispersal of the chemical elements in the galaxy—the very chemical elements which make up the Earth and make life on Earth possible. Nucleosynthesis is the study of the origin of the chemical elements. One might think that all of the elements of the universe were made when the universe itself was created. That is not the case at all, and observations of the oldest stars and models of the birth of the universe indicate that only hydrogen and helium were made when the universe was created. The heavier elements, including those which make up most of the Earth and what is in our bodies, were made in stars; many were made during supernova explosions, and the supernova explosions are what spread the newly—made elements into the interstellar medium so that new stars could be formed from the enriched material. Thus, astronomers are interested in more than the answer to the question of what a supernova explosion is. They are interested in how it affects the galactic environment as well.

Although mankind has seen supernovae for millennia; we only know about a few seen since history began to be recorded. Supernovae can be confused with other events, including the passage of bright comets and the less violent stellar explosions called novae. The older historical records do not distinguish between novae and supernovae; we have to do that after the fact. Both novae and supernovae brighten very quickly, and at maximum novae become about a hundred—thousand times brighter than the Sun, whereas supernovae become up to a thousand—million times brighter than the Sun (if seen from the same distance). "Very quickly" means within about 24 hours: this distinguishes novae and supernovae from comets, which take months to brighten very much. Therefore, from the point of view of someone observing with the naked eye, a star would appear suddenly where there was no star the night before, would stay there awhile (months), and gradually fade away into invisibility. Thus they were dubbed "guest stars," "temporary stars," or "new stars." The last of these, translated into Latin, the universal language used by scientists in Europe in the middle ages and post—renaissance period, was "nova stella." At some point it became customary to drop the "stella," promote the adjective to a noun, and refer to the "nova stella" as a "nova." The word "nova" refers to one of them, the word "novae" (the Latin plural) refers to more than one.

We will discuss (in Chapter 3) the six bright supernovae seen in 185, 1006, 1054, 1181 1572 and 1604 A.D. All became extremely bright; all occurred before modern technology could be used to study them. But we know something about these explosions besides what we find in the historical record. When such a titanic explosion occurs, the debris is blown outward at fantastic speeds. We see the remains a thousand years later as an expanding cloud of gas—a nebula—called a supernova remnant. The most famous of these is the Crab Nebula in the constellation of Taurus; it is the remnant of the supernova of 1054, and the subject of Chapter 4. Another famous remnant is the Cygnus Loop, containing the Veil Nebula; it is the remnant of a supernova which exploded twenty or thirty thousand years ago. Another old remnant is the Gum Nebula, an enormous bright filamentary ring apparently surrounding the Vela supernova remnant. Except for these three nebulae found with optical telescopes, the remains of supernova explosions are usually found through observations with radio telescopes. To date about 135 remnants have been found in our galaxy. Nebulae are not the only kind of remnant of supernovae; some supernovae produce pulsars, and a few may produce black holes. A few pulsars are found in the nebular remnants of supernovae. Several hundred others are known which are not associated with nebulae, but which are thought, for good reasons, to have been produced by supernovae. We will discuss the nebular and stellar remnants of supernovae in Chapter 11.

As noted above, supernovae are understood to be the death throes of certain stars. The study of supernovae includes attempts to model theoretically the details of the explosion and its aftermath. In order to understand the starting conditions of the explosion, the evolution of the star up to the beginning of the explosion must be understood. Mankind has wondered about the nature of stars since time immemorial. But it has been only in the past century and a half that any real progress has been made in understanding their physical structure and evolution. We now know that most of the stars one can see, including our Sun, have in their seething cores a powerful nuclear reactor which generates the energy to make the whole star shine for millions or thousands of millions of years. Although it is the same nuclear reaction which makes a hydrogen bomb explode, the weight of the stellar mass surrounding the core contains this ongoing explosion and lets it continue in secure equilibrium for almost its entire lifetime. Even though the use of the star's nuclear fuel is so efficient that it may last for thousands of millions of years, this slow conversion of one nuclear species into another in the central regions of the star will produce changes in its entire structure. These structural changes are what we call the evolution of the star.

The study of stars is complicated by the fact that we must study the interior structure by what we see from the outside. This means that we must rely on theoretical models. A certain interior structure implies a certain external appearance; matching a model to the external appearance implies that the interior of the real star is well described by the model. The models depend upon a knowledge of physical laws applied to the conditions found in the material of the star. Thus, the progress of the models has depended upon our progress in physics. Since the models are computed with the largest and fastest computers available, their progress also has depended upon advances in computer technology.

The evolution of stars, from their formation out of gigantic clouds of gas and dust in interstellar space to the stages when they are ripe for the supernova explosion, is now fairly well understood by astronomers. We will sketch our present knowledge of the nature of stars and describe the evolution of a star from birth to death. The supernova explosion, itself, is reasonably well understood. But the appearance of a bright supernova presents an opportunity to test the models in ways not possible with the fainter ones. We will also describe models of supernova explosions.

But this book is about the supernova which we first saw in February, 1987 in the Large Magellanic Cloud. Once the discovery had been announced to the world, the fun began! Astronomers began observing the supernova intensely. Three circumstances made the effort of observing the phenomenon challenging. One was the fact that it could be seen only from the Earth's southern hemisphere. This is significant because most of the astronomers and telescopes are in the northern hemisphere. Another was the fact that, as nice as it was to have a bright supernova to observe, it was too bright for many telescopes and instruments. This results from the fact that astronomers usually are observing faint stars and galaxies. Much progress in recent years has come in the observing of fainter and fainter objects. The instruments have become more sensitive, and have become less and less able to observe bright objects. But astronomers are versatile, and so are their instruments. They quickly found ways to adapt the instruments to the situation, and the race was on to gather information as rapidly as possible. The third circumstance which made the observing of the supernova challenging was that the telescopes had already been scheduled to make other observations. Telescope time is precious, and essentially all observing is for the study of objects which do not appear suddenly. In the typical case, astronomers plan their observations months or years ahead, and telescopes are scheduled for months before the nights of observation. What happens when something like Supernova 1987A occurs? We'll tell the story of what happened in Chapter 13.

This supernova also has the distinction of launching a new branch of observational astronomy. For the first time, neutrinos were detected which had originated in an object outside our own solar system. Previously, only the Sun was observed to be a source of neutrinos, and even in this case, the observations were difficult to understand. Those who concern themselves with the study of neutrinos got a gratifying result.   After the embarrassment of having the rate of detection of neutrinos from such a well—understood star as the Sun disagree with the models, they found for the supernova just what they had predicted.   The neutrino observations are important because they tell us some things about the supernova which we can't find in any other way, and they may tell us something about the neutrinos themselves: their mass.

Almost immediately after the announcement of the discovery of 1987A, astronomers who specialize in theoretical models of supernovae got to work.   As the initial observations came in, and both the typical and unusual aspects of this supernova became apparent, the models were adjusted to fit.   It became apparent that the star which was identified as the precursor was not what had usually been assumed to be the precursor of a "Type II" supernova.   Nevertheless, the models were able to accommodate the unusual precursor.   The modeling of supernovae is an active field of astronomy, and we tell about some of the participants and their models in Chapter 15.

Although they often work alone, astronomers are also social beings. When something exciting happens, they are especially prone to get together to talk about it.   National and International meetings to discuss research results are common, and a number of such meetings which had been scheduled long before the supernova occurred became impromptu supernova sessions.   A typical and important example was the joint meeting of the American Astronomical Society and the Canadian Astronomical Society, held in Vancouver, BC, on June 14–18, 1987. Many of the experts on supernovae were present, and the latest results were presented, both observational and theoretical.   Furthermore, the astronomical societies held a press conference.   The excitement was intense, and the story is told in Chapter 16.

As noted above, we see supernovae commonly in other galaxies. Supernovae are estimated to occur once every 20 to 40 years in the brightest galaxies, which means that we have to look at a lot of galaxies to detect one.   In the past 50 years, about 500 have been discovered. This reflects the systematic search in recent years by a number of astronomers—it hasn't happened by accident.   Currently, three approaches seem to be taken.   One, exemplified by the Australian amateur astronomer Robert Evans, is to observe visually many galaxies each night. He can compare what he sees with photographs and charts, and can recognize when a new star has made an appearance.   After years of observation, he has "memorized" the appearance of a large number of galaxies, and that speeds up the process.   He has personally discovered over a dozen supernovae.   A second approach was pioneered by Fritz Zwicky, and has been followed by many others since: To photograph many galaxies repeatedly night by night, and compare the photographs to find "new stars".   The disadvantage of this method for professional astronomers is the large amount of time required.   For amateur astronomers the disadvantage of this method is the expense of building and operating a suitable telescope and the expense of the photographic materials.   The third method is to construct an automatic telescope system which does the observations with minimal human intervention.

The observations are recorded in a form which can be entered into a computer. At that stage, one can search the images visually or have the computer do it automatically. Each approach is in use and we tell their stories.

As exciting as 1987A is, we are still waiting for the long—overdue bright supernova in our own galaxy. What kind of star do we expect to be the victim, and what would the effects be on the Earth if one "goes off" in our (celestial) back yard? It certainly could be very bright, and that could be a real problem. But the great brightness would present opportunities for observations which are not even possible for Supernova 1987A. And, as will be the case with 1987A, the development of the remnant during the years after the explosion would be one of the most exciting opportunities for astronomy. We will speculate on the occasion of such an event in Chapter 17.

Our goal in this book is to describe in detail what supernovae are and why they are important. We tell the exciting story of the discovery of Supernova 1987A, and how the community of astronomers has reacted to the discovery. It is a great adventure!

CHAPTER 3

HISTORICAL SUPERNOVAE

Much of our knowledge of supernovae in our galaxy comes from ancient records of naked eye astronomical observations. It is not easy to make sense of these records. One must deal with diverse languages, foreign cultures, and even unique calendars. There are gaps, duplications, contradictions, inconsistencies, and inaccuracies. And there is the additional problem of distinguishing between the novae and supernovae observed long before it was known that they are different phenomena. Nevertheless there is a consensus now in the astronomical community that over a half dozen supernova explosions in our galaxy were observed, fixed in time, pin—pointed in the sky, and recorded permanently in historical chronicles. These are listed in Table I.

TABLE I

The Historical Supernovae

Date When Brightest	Magnitude	Constellation	Duration	
1. 185 A.D.	−8	Centaurus	20	months
2. 1006	−9.5	Lupus	30	
3. 1054	−5	Taurus	22	
4. 1181	0	Cassiopeia	6	
5. 1572	−4	Cassiopeia	18	
6. 1604	−3	Ophiuchus	12	

The earliest recorded sighting was in 1112 B.C., inscribed on animal bone by ancient Chinese observers. This sighting is not included in Table I because most of the details are quite sketchy compared to the others: the date was not precise, there was no indication of its location, brightness, or duration, and we cannot really know for sure that it was a supernova rather than a nova or perhaps even a bright comet.

No doubt the transient bright star in the sky for which many would like an explanation is the Christmas star, or the Star of Bethlehem. Much effort has gone into trying to find a satisfactory astronomical explanation of the Biblical story. Since it occurred almost two millennia ago, historical records are understandably sketchy and difficult to rely upon literally. Much of the available historical evidence actually comes from the Bible itself. Astronomical explanations have included suggestions that it was a striking conjunction of two or possibly three of the brighter planets, that it was a bright comet, or that it was a nova or supernova. Since this book concerns supernovae, let us comment on just the last of these three possibilities. One historical record, from an ancient Chinese source, refers to a bright object seen in March or April of the year 5 B.C. in the constellation we call Capricorn (the Sea Goat) and remaining visible in the night sky for more than 70 days. How consistent is this with historical record? The year is consistent because thorough analysis of ancient calendars has shown that Jesus was actually born sometime between 7 B.C. and 5 B.C. The position of Capricorn in the spring sky as seen in the near east is consistent. The time of year itself is consistent because, after all, the shepherds were tending their flocks at the time of the birth. And the 70–day duration would have been enough time for the wise men to have made their journey, having it guide them as they went. It was not, however, as bright as Biblical accounts describe it. And a stellar object would not have executed the odd movements in the sky which the Bible seems to suggest. Little more can we say conclusively. In any event, whether or not the bright star in 5 B.C. was the Star of Bethlehem, the relatively short duration of visibility suggests it was a nova rather than a supernova.

The earliest known definite sighting was the supernova of 185 A.D., which occurred in the constellation we now know as Centaurus (the Centaur, the half–man–half–horse). The only scientifically useful records of this supernova were made by Chinese astronomers, who were consistent observers of various astronomical phenomena throughout the millennium before the birth of Christ. It can be reconstructed that this particular one was first spotted on the night of December 7th, reached a maximum brightness of −8th magnitude, and remained visible to the naked eye for 20 months. There is mention in a couple of Roman chronicles of what probably was the 185 A.D. supernova, although no information is provided regarding its position in the sky and in only one of them can the year be fixed as probably 185 A.D. In that one, however, it is useful that it mentions the star shining continuously by day. At magnitude −8, the 185 A.D. supernova definitely could have been seen in broad daylight.

Clearly the most dramatic of the historical supernovae must have been the one which occurred in 1006 A.D. in the constellation we now know as Lupus (the Wolf). This date should not be confused with 1066 A.D., which was the year of the Norman invasion of England, and also a year in which Halley's comet swept past the Earth on its way around the Sun.

Astronomers scrutinizing ancient records have been able to deduce that the 1006 supernova was first noticed in Japan on the night of April 28th (possibly), or the night of April 30th (probably) by the 26—year—old Egyptian astronomer Ali ibn Ridwan, living in Cairo, and on the night of May 1st (definitely) by several different Chinese astronomers, and on the night of May 3rd (definitely) in what is now Iraq. The Chinese observations are particularly useful in attesting to the fact that it became even brighter after its initial detection. It eventually reached a peak brightness of −9.5 magnitude, which corresponds to the brightness of the half—illuminated Moon, i.e., the Moon at quarter phase. This is much brighter than any star or even any planet. Our sister planet Venus at its brightest reaches magnitude −4.3 (and is responsible for more "U.F.O." sightings than any other astronomical object). The supernova in Lupus became 100 times brighter than Venus ever does and thus must have been visible in the daytime for months. It is certain that it continued to be visible at night for at least two and a half years, and possibly for several years, before fading from sight. Altogether, it was seen and recorded on three continents. In addition to Japan, China, and Egypt, there were sightings in other countries of the near east, in northern Africa, and in the three European countries of Italy (three different locations), France (two different locations), and Switzerland (two different locations).

Perhaps the best known, most intensely studied, and most popular historical supernova is the one which gave rise to the Crab Nebula (which we describe in detail in Chapter 4). Although it occurred in 1054, soon after the 1006 supernova, no record of its appearance has ever been found anywhere in Europe. Its occurrence in the constellation of Taurus (the Bull) made it ideal for viewing high in the nighttime sky in Europe, and it took place in the lifetime of many who must have seen the 1006 supernova. It was, however, documented very well in the Orient. Chinese records show it first being visible to the naked eye on July 4th, reaching a peak brightness of magnitude −5, being visible in the daytime for 23 days, and fading from view only after 22 months. Its appearance was also recorded extensively in Japan. There was one possible report from the Middle East, by the Christian physician Ibn Butlan, from Bagdad, who blamed it for a major plague (probably bubonic) which killed thousands in Constantinople, Cairo, and in many cities in Iran, Yemen, and Spain in that same year. Sadly, there is no record of Ali ibn Ridwan, discoverer of the great 1006 supernova, ever having seen this 1054 supernova, even though he didn't die until 1061.

There are drawings on cave and canyon walls by American Indians of the Southwest which might correspond to its sighting in the western hemisphere. Altogether there are 3 in Arizona, 6 in Nevada, 5 in New Mexico, 1 in Texas, 1 in Utah, 4 in California, and one in Baja California, Mexico. None of these pinpoint the exact location in the sky nor fix the event accurately in time. In some of them, the bright star is depicted as being the same size as (presumably they meant to indicate it was the same brightness as) and very nearby the Moon. If we assume the year was indeed 1054 A.D. and that the location was indeed the supernova's exact position in the constellation of Taurus, then the relative positions of the star and the Moon tell us that the date corresponded to the morning of July 5, which would be the evening of July 4th. This is the same night on which the Chinese recorded the first definite sighting. In one case, that of the cave paintings in Navajo Canyon, Arizona, we know that Indians were living in that location at a time somewhere between AD 700 and AD 1300, consistent with a 1054 A.D. sighting. Many people don't realize that Venus at its maximum brightness, −4.3 mag., can be seen in the daytime, though with some difficulty. Thus the 1054 supernova at its maximum brightness, −5.0 mag., was two times brighter and thus would have been easier to see in the daytime than Venus. At night, it was the brightest object in the sky except for the Moon. Europe must have been "sleeping" at the time, and indeed it was, culturally and scientifically, compared to the Renaissance which began a half millennium later.

The third most recent historical supernova occurred in 1181 A.D. in the constellation of Cassiopeia (the Queen on her Throne), but not many people are familiar with it. Unlike Halley's comet, which comes by Earth every 76 years and lets some people see it twice in a lifetime, it is doubtful that any 130−year−olds were living in 1181 to make the comparison between it and the more conspicuous 1054 outburst. The 1181 supernova reached only magnitude 0 at its brightest, but this means that there were only a few stars brighter than it in the northern half of the sky. Its sighting was recorded only in China and Japan. There is no trace of sightings in Korea, the mideast, Europe, or America. And it did not leave a prominent remnant, as did the 1054 supernova, which left the spectacular Crab Nebula. The first observation was made on August 6th in southern China and the second on the very next night, August 7th, in Japan. The last recorded sighting pegged the duration of nighttime visibility at 185 days.

The world had to wait almost four centuries for the next supernova to be seen, but this one was a good one, almost as bright as the one in 1054. At maximum it reached magnitude −4, only a shade fainter than Venus at its very brightest. The first definite sighting, with the time and date firmly fixed, was by Wolfgang Schuler of Wittenberg (Germany) at 6 a.m. on the morning of November 6, 1572. A possible earlier sighting, though the date was specified somewhat ambiguously, was by Francesco Maurolyco, a mathematician in the abbey at Messina (Sicily). In the history of supernovae we have a first: a date when it was certain the supernova was not visible in the night sky. Jerome Munoz, a mathematician and professor of Hebrew at the University of Valencia (Spain) was sure he did not see it on the night of November 2. There were recorded sightings also in China and Korea. The Chinese first saw it on November 8 and the Koreans specified its first appearance only to the nearest month. The overall duration of visibility was 18 months, almost as long as the 1054 supernova. Although not the discoverer of this supernova, the famous Danish astronomer Tycho Brahe took it very seriously and made numerous detailed estimates of its brightness and also extensive accurate measurements of its position in the sky. He used these position measurements to prove that it was not like the planets or comets (which move around in the sky relative to the stars) but more like the stars (whose relative positions are fixed firmly in the sky). To this day we refer to this 1572 event as Tycho's Supernova. The chronicle of this supernova is sufficiently complete that we can plot its "light curve," a detailed representation of its brightness change with time (See Figure 1).

The most recent supernova observed in our galaxy exploded in 1604, only one generation after Tycho's supernova of 1572. That was, ironically, just 4 years before the telescope was invented and only 5 years before the first astronomical observations made with a telescope, by Galileo Galilei in 1609 in Padua. No supernova in our own galaxy has been observed or studied with the aid of a telescope. Although commonly known as Kepler's Supernova, the first sighting was actually made on the night of October 9th simultaneously and independently by a medieval physician (to this day we don't know his name) and the astronomer I. Altobelli, both in Italy. The famous observer Fabricius, whom we have mentioned before, testified that he was sure he did not see it on the night of October 8th, the very night before. The great German astronomer Johannes Kepler, known for the three laws of planetary motion which helped lead Sir Isaac Newton to his Law of Gravity, did observe the supernova keenly and consistently. Unfortunately, Tycho Brahe could not see this fine supernova, because he had died in 1601, just 3 years before its outburst. Galileo did, however, observe it, although not as extensively as did Kepler. It reached magnitude −3.0 at peak brilliance and was sighted for the last time almost exactly one year after its initial discovery. There were other sightings elsewhere in the world: China, Korea, and at other places in Europe. The light curve of this last−seen galactic supernova is also shown in Figure 1. This light curve is based on the best observations, *i.e.*, those made by the Europeans and the Koreans. Note that, in this one rare instance, it is clear that the first sighting occurred well before the supernova reached its maximum brightness. This last supernova seen in our galaxy is commonly referred to as "Kepler's Supernova." We cannot explain (and have never heard the question raised) why it was Kepler's (Johannes Kepler's last name) Supernova in 1604 but was Tycho's (Tycho Brahe's first name) Supernova in 1572. Perhaps it is because Tycho was of noble birth, and Kepler was a commoner.

Observations of other galaxies suggest that a supernova explodes roughly every 30 years in a typical galaxy, and ours is a typical galaxy. The long interval of 383 years that has elapsed between 1604 and 1987 suggests that either something is wrong with the calculations or that we are missing most of those which do occur. The latter is probably the case, for reasons we will explain next.

The "empty space" between the several hundred thousand−million stars which make up our galaxy is not empty. Interstellar space contains a very tenuous gas (molecules, atoms, and ions) and some dust (small particles or grains of matter). The interstellar dust, which astronomers think is like soot and rock dust covered with a layer of ice, scatters light. There isn't much gas and dust in a given volume of interstellar space, but the distances across the galaxy are so great that the dust can significantly dim starlight.   Therefore, just as you can't see well across a very smoky room or down a very dusty road, you cannot see stars far away across a dusty galaxy.   To illustrate how severe this interstellar extinction can be, consider a star on the opposite side of our galaxy from where we are situated, about two thirds of the way from the galaxy center.   It would appear only 0.000001% as bright as if it were viewed through perfectly transparent interstellar space.   It is by this and related arguments that astronomers have convinced themselves that supernovae probably do occur in our galaxy at the three−per−century rate expected theoretically.   We are unable to see most of them.

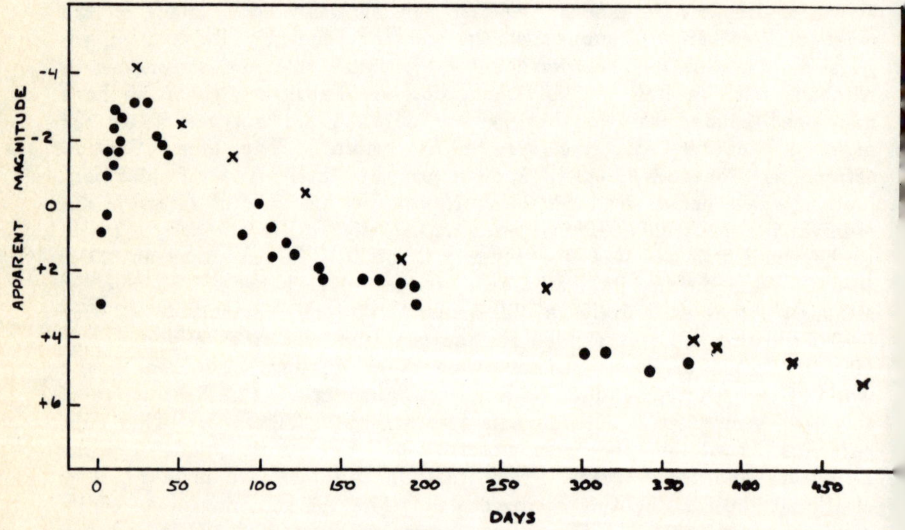

Figure 1.   The combined light curve from the observations of Tycho's (1572) and Kepler's (1604) supernovae have been converted to the modern scale of apparent magnitude and are shown plotted against time.   The X's refer to Tycho's and the dots refer to Kepler's.

That is probably part of the explanation of what happened in the case of the supernova of 1667, the supernova which happened but was never seen.   It may have been intrinsically fainter than the typical supernova, but it is in a region heavily obscured by interstellar dust.   It would have been the most recent supernova of all those listed in Table I, if anyone had seen it.   Astronomers know of its existence and the approximate date of its occurrence as a result of radio telescope observations (see Chapter 4).   As a radio source it is known as Cas A, the first discovered (in 1948) in the constellation Cassiopeia.   (See Figure 2.)   It is also the most intense radio source anywhere in the sky, except for the Sun.   Since then, astronomers have obtained photographs of the circular nebula characteristic of supernova remnants (the expanding cloud of gas produced by the explosion).   The determination the of expansion age (see below and Chapter 4) gives the time of the explosion as sometime around 1658, give or take 3 years.   Subsequent estimates, trying to take into account the expected slowing down of the expansion rate, place the occurrence closer to 1667.   A few years ago it was suggested that the explosion had been seen after all.   A 5th magnitude star (denoted 3 Cas) is plotted on a star chart constructed from observations made on April 16, 1680 by the first Astronomer Royal, George Flamsteed. On a later star chart of that same region, 3 Cas is not there and no other star is near its place.   The approximate coincidence in time and location makes it tempting to think Flamsteed did (unknowingly) witness and record the event which gave rise to the radio source Cas A, but subsequent scrutiny of the argument reveals that the times and locations are not really close enough to make the case compelling.   A star in our galaxy did explode as a supernova on or about 1667, that much is certain.   Whether or not anyone living at the time saw it is not certain.

Until only a few decades ago, no one knew what novae and supernovae really were, apart from their appearance as a bright new star in the night sky.   They were lumped together as the same phenomenon: a bright star appears one night where apparently no star had been the night before.   The first breakthrough in understanding these phenomena came in the very 1900's, when astronomers first took photographs of the spectra of Nova Persei 1901, when it was still near maximum brightness at about zeroth magnitude.

The spectrum of Nova Persei showed dark (absorption) lines during the time of declining light.   It was possible for the astronomers to use their knowledge of the Doppler effect to determine that the gas blown out in the explosion was expanding outwards from the center at extremely high speed—1500 kilometers per second! The Doppler effect makes this determination possible, because it is a relative shift in the position of an emission line as a result of the motion of the source towards or away from us (see Figure 3).   Thus, they verified that they were seeing a real explosion.   Dramatic confirmation was found on photographs taken two decades after the explosion.   The star had faded far below naked eye visibility, and a small circular nebulosity was present.   The ejected gas had spread out in all directions and grown to an extent where it could be seen and photographed.   When the size of the nebula was measured, and with the distance and the time of the expansion known, it was possible to determine the rate, or velocity, of expansion.   It was the same velocity as measured by the Doppler effect; they agreed well.   In this case the distance had been determined independently, but for most novae and supernovae the determination of the distance can be done only approximately, if at all.

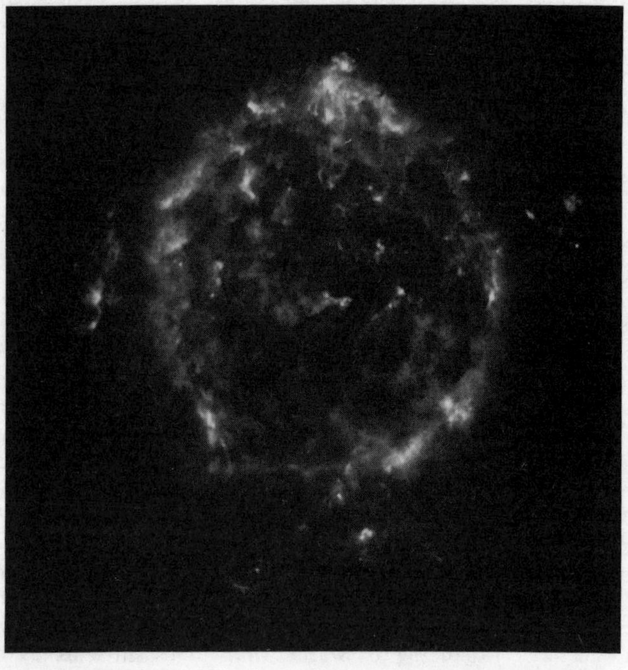

Figure 2.   The supernova remnant, Cassiopeia A, at radio wavelengths.

A second breakthrough was made when a nova appeared at precisely the same position of a previous nova of decades before. That told astronomers that a nova is not a "new star;" rather it is an explosion of some kind experienced by an already—existing star. A repeat performer like this is called a "recurrent nova," and the first known recurrent nova was T Pyx, which blew up in 1890 and again in 1902, 1920, 1944, and 1967. This many recurrences is unusual; in a recent listing, we find 3 novae which have been observed to explode twice, one which has exploded four times, and one (T Pyx) which has exploded five times.

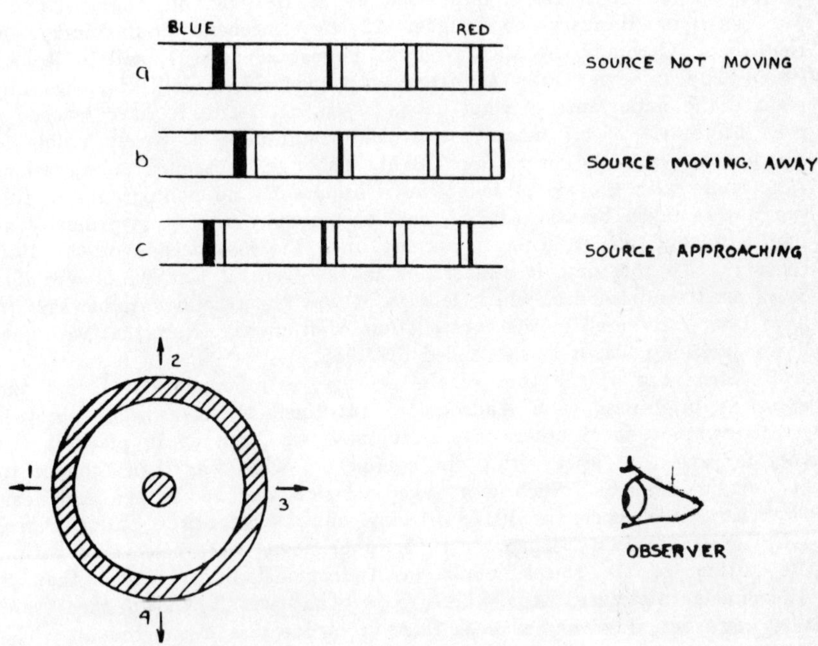

Figure 3.   The Doppler effect.   In a) the spectrum of the source is shown with bright lines.   The source is not moving towards or away from the observer.   In b) the spectrum is shown for a source moving away from the observer.   The lines are shifted towards the red.   In c), the spectrum is shown for a source approaching the observer.   The lines are shifted towards the blue.   In the lower part of the figure, an expanding shell of gas is shown.   The observer sees a spectrum which includes contributions from each part of the shell.   The spectra from parts (2) and (4) would be unshifted, as in a), above, since the motions of these parts of the shell are crosswise and not towards or away from the observer.   The spectrum from part (1) would be shifted towards the red, as in b), above.   The spectrum from part (3) would be shifted towards the blue, as in (c), above.

The recognition of recurrent novae, however, still did not give astronomers even an inkling that supernovae were different from and spectacularly more significant than novae. One particular supernova played a pivotal role in the eventual realization of the difference. That was the variable star discovered in 1885 near the center of M31 (also known as the great nebula in Andromeda). The first actual sighting was at the Dorpat Observatory, in Russia, by E. Hartwig, on August 20. It was seen in Hungary on August 22 by Baroness Podmaniczky, in Heidelberg, Germany, by Max Wolf on August 25 and 27, and in Rouen, France, by Ludovic Gully as early as August 17. Only Hartwig fully realized the importance of what he had seen. It seems to have reached a peak brightness of 5.7 magnitude, which would make it barely visible to the naked eye on a clear, dark night, although subsequent observations were made with the aid of telescopes. Apparently no photographs of this event were taken because astronomical photography was quite primitive at that time and didn't begin in earnest until the last decade of the 19th century. To this day, it still stands in the General Catalog of Variable Stars as S Andromedae, which tells us it was the second variable star to have been discovered in the constellation Andromeda. Now that we know it was a supernova, it is also called SN 1885.

Shortly after the turn of the century, astronomers noted that the apparent brightness of S Andromedae (assumed to have been a nova), compared to that of other nova outbursts, was such as to place it not very far away, at least within the confines of what was then thought to be our own galaxy. Such was the conclusion reached by the American physicist F. W. Very in 1911, drawing on observations of Nova Persei 1901 for comparison. Then, in 1917, other novae were discovered within the outline of the spiral nebula in Andromeda, much fainter than S Andromedae (because, as we know now what wasn't known then, that they were actual novae, whereas S Andromedae was a supernova). The faintness of these true novae, properly identified as such, was used by Heber D. Curtis in his Great Debate with Harlow U. Shapley on April 6, 1920, at the National Academy of Sciences in Washington, D.C., to argue that they must be so extremely far away as to prove that the spiral nebula in Andromeda cannot be within our galaxy but must be a galaxy outside our own. That argument was only one of several used by him, against counter−arguments made by Shapley that the spiral nebulae are within our own galaxy. It is ironic that, if Curtis was right (and we know today he definitely was), then S Andromedae must have been an incredibly bright nova the likes of which could hardly have been imagined, and certainly was not understood or explained. The Great Debate ended as a stalemate as far as persuading the majority of scientific opinion.

The real breakthrough didn't come until three years later. In 1923, the American astronomer Edwin Hubble, taking long exposure photographs with the 2.5−meter telescope atop Mt. Wilson near Pasadena, California, discovered pulsating variable stars in M31 and some other nearby spiral nebulae. He found them to be so extremely faint that they had to be very far away, so far away that the spiral nebulae had to be outside our own galaxy, hence galaxies in their own right. That finally, belatedly, but decisively settled the Great Debate. Not long afterwards astronomers recalled what had happened in 1885 and realized that, because it was now certain this 6th magnitude outburst had originated so far away (M31 is about 2 million light years from Earth), S Andromedae must have been a uniquely superluminous nova outburst. Hubble himself, in 1929, realized this. Meanwhile, examination of photographs of other spiral nebulae (now called spiral galaxies) had revealed other examples of "novae" comparable in brightness to the galaxies in which they occurred, *i.e.*, not like the fainter outbursts which had been noted in M31, which were bona fide novae.   In 1925, the Swedish astronomer K. Lundmark (1889–1958) distinguished clearly between the "upper–class" and "lower–class" novae. It was Fritz Zwicky (1898–1974), a Swiss–American astronomer working at the California Institute of Technology, who first coined the term "supernova."   He first used the term "super–nova" during one of his lectures at Caltech in 1931 and first used the term officially at the December 1938 meeting of the American Physical Society.

From that point on, astronomers have been investing considerable time and effort photographing galaxies, carefully comparing photographs of the same galaxy taken at two different times, and trying to discover supernova explosions in progress.   Finds such as these have given us much information, but they are not nearly as exciting as a bright, nearby supernova in our own galaxy.

FRITZ ZWICKY

Fritz Zwicky (1898–1974) was one of the pioneers in the understanding of supernovae.   His brilliant speculation on the nature of the supernova explosion was combined with years of patient observations designed to discover supernovae in other galaxies.   Zwicky was born in Bulgaria of Swiss parents.   He was educated in Switzerland, receiving his PhD in 1922 from the Swiss Federal Institute of Technology in Zurich.   He went to the California Institute of Technology, in Pasadena, California, on a fellowship in 1925.   He became an assistant professor in 1927, and stayed for the rest of his life, rising through the ranks to full professor.   He kept his Swiss citizenship for his entire life, even though he stayed through retirement at Caltech and died in the United States.

His inspired speculation on the supernova explosion and its relation to the formation of neutron stars resulted from a collaboration with Walter Baade in the 1930's.   During the 1930's he also began his surveys to find supernovae in other galaxies.   It is said that his first attempts to survey the Virgo cluster of galaxies from the roof of the laboratory at Caltech with a small camera were met by the great mirth of his colleagues, but even without success he was able to convince G. E. Hale to fund the construction of an 0.45–meter Schmidt telescope (a wide–angle camera) on Palomar Mountain.   With the aid of the 0.45–meter from 1936 to 1958 (with a gap during WW II), and the 1.2–meter Schmidt from 1958 to his death, Zwicky personally found 122 supernovae!

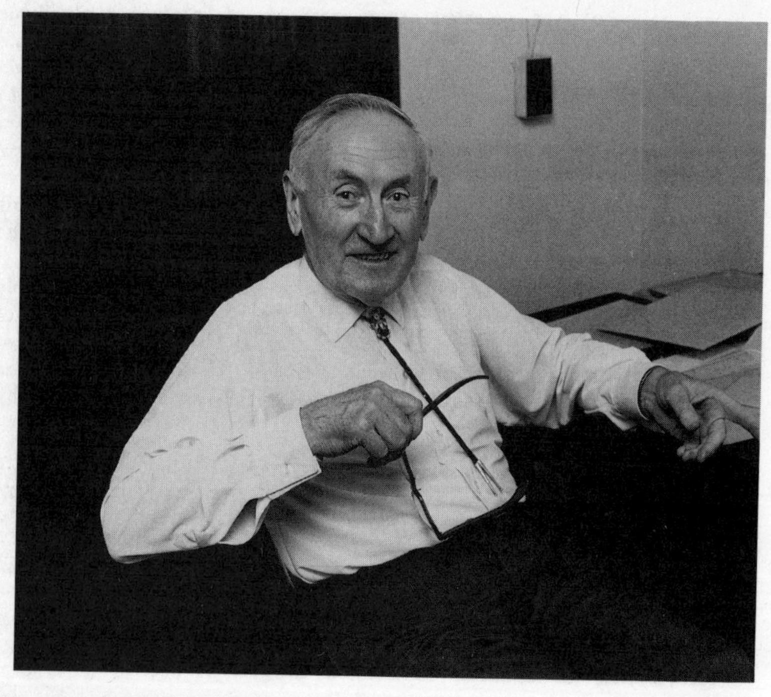

Fred Zwicky, supernova research pioneer.

CHAPTER 4

THE CRAB NEBULA

THE DISCOVERY

The Crab Nebula is thought to be the remnant of the supernova of 1054 A. D., for reasons which we will see in this chapter. It is by far the best known, and is the brightest supernova remnant. It has been studied in more ways and in greater detail than any other remnant. Let us, therefore, give a step—by—step chronology of the many significant facts which have been learned about it.

The nebula was first seen with a telescope by the English physician and amateur astronomer John Bevis and appeared in a sky atlas which he prepared in 1745 or thereabouts. In 1758, the French astronomer Charles Messier discovered it independently, while observing a comet. In 1771 he published a list of 109 easily visible nebulae. The term nebula, the Latin word for cloud, used to denote any fuzzy patch of light in the sky; we now know it includes other types of astronomical objects besides supernova remnants including emission nebulae, reflection nebulae, planetary nebulae, and spiral nebulae. All but the last of these are objects found within a galaxy; the spiral nebula are galaxies themselves, similar to our own but far beyond the stars on the outer boundaries of our own. Messier's interest was not, ironically, in the nebulae themselves. As a noted comet hunter of his day, his purpose was to provide a list of fuzzy objects, always present at fixed positions in the sky, which might be mistaken for slowly moving comets. During the 18th and 19th centuries, the discovery of comets was one of the most popular pursuits for astronomers, both amateur and professional, and it was quite an embarrassment to announce discovery of a new comet only to have it prove to be "merely" one of those other fuzzy objects which are always there. It is an interesting coincidence that the first nebula on Messier's list, commonly referred to as Messier 1 or M1, is the Crab Nebula.

Starting with William Herschel, nineteenth—century observers began to use larger and larger reflecting telescopes. With even the largest of these, M1 remained a nebula, although its appearance suggested that it was near being resolved into faint stars. It was observed by William Parsons, the 3rd Earl of Rosse, in Ireland in 1844. In that year he published a picture of the nebula, drawn at the telescope, which suggested the name "Crab Nebula." The picture looks somewhat crab—like, but doesn't look very much like the nebula. A later drawing, made using Lord Rosse's telescope and published in 1879, looks much more like later photographs.

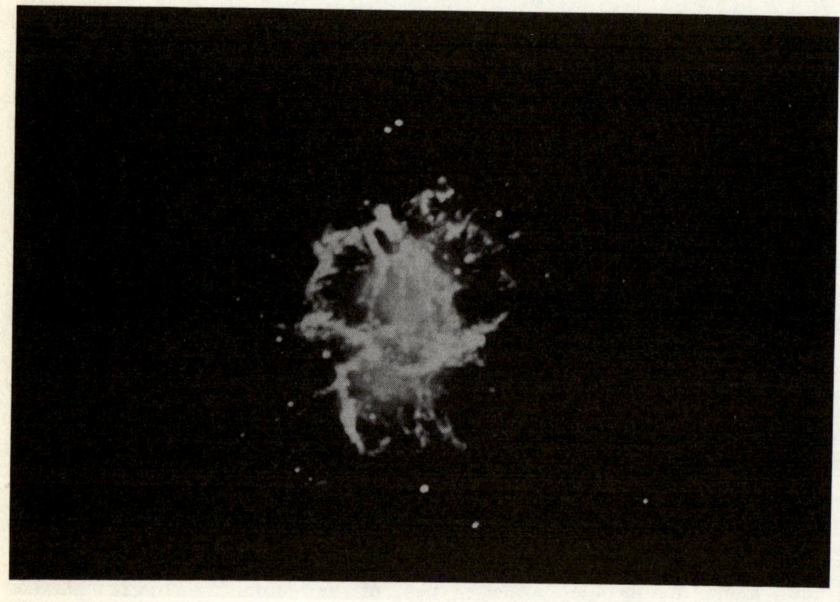

Figure 1.   The Crab Nebula photographed in visible light (Courtesy NOAO).

The spectrum of the Crab Nebula was first observed visually by Winlock and Pickering at Harvard in 1868. They saw bright emission lines known to be characteristic of gaseous nebulae. They also saw a continuum, which means that they saw a complete spectrum, including light with all wavelengths within the interval covered by their spectra. The common sources of continuous spectra known at that time included stars, which made their observations very confusing. The spectrum was first photographed in 1913 by the American astronomer Vesto M. Slipher at Lowell Observatory in Flagstaff, Arizona. He was able to measure the positions of the spectrum lines very precisely and, by invoking the principle of the Doppler effect, he deduced that it was expanding in all directions at approximately 900 kilometers per second.

In 1921, American astronomer John Duncan compared a photograph he had taken with an earlier (1919) photograph taken by G. W. Ritchey using the same 1.5–meter reflector at Mt. Wilson. From the comparison, he concluded that the Crab Nebula was growing in size, expanding in all directions from some point in the center. In that same year, 1921, Swedish astronomer Knut Lundmark published one of the first lists of novae (the distinction between novae and supernovae wasn't known at that time) observed by ancient Chinese astronomers and noted that, in approximate position, the one sighted in 1054 coincided with the Crab Nebula.

In 1928, the American astronomer Edwin Hubble took the decisive step of comparing Duncan's expansion rate with the current size of the Crab Nebula and, by "turning back the clock," determined that the expansion had originated from a central point in the year 1075, give or take about 50 years. Thus Hubble was the first to point out that the Crab Nebula and the 1054 supernova were closely coincident in age as well as in position. Unfortunately, because he didn't publish his conclusion in the regular astronomical literature, it was overlooked by other professional astronomers and had to be rediscovered. This was done by astronomer Jan Oort and oriental historian J. J. L. Duyvendak, both from the Netherlands, and their work became known in 1942.

Also in 1942, the American astronomers Walter Baade and Rudolf Minkowski considered the apparent motions of the two stars of similar brightness lying on a NE–SW line near the center of the nebula. They projected these motions back 900 years, and compared the resulting positions with the center of expansion of the nebula. They identified the 14th magnitude "southwest star" as having been at the true center of the expansion; they also were the first to suspect that it was still interacting with the nebula.

The distance to the Crab Nebula was first determined by the American astronomer N. U. Mayall at Lick Observatory. His method was to compare the apparent angular size of the nebula with the true size deduced from the age and the velocity of expansion measured from the Doppler shift. We now adopt a distance of between 5,000 and 6,000 light years, making it a relatively close object on the scale of the galaxy (about 30 times closer than SN 1987A). With the distance known, we can calculate the intrinsic maximum brightness of the supernova of 1054; the latest estimates indicate that it was about 500 million times as bright as our Sun. This value of the maximum brightness makes it pretty typical of a Type II supernova (see Chapter 6).

The two principle pieces of evidence that the Crab Nebula is the remnant of the supernova of 1054, the coincidence in position and the coincidence in time of origin are not without controversy. The Chinese record clearly states that the supernova was located to the southeast of the star Zeta Tauri. The Crab Nebula is located to the northwest of Zeta Tauri. The Chinese records are generally our more reliable sources of information about historical supernovae, so this discrepancy might seem to be serious. Adding to the controversy is a discrepancy in the apparent time of origin of the nebula. If we calculate the time of origin of the nebula, knowing how large it is and how rapidly it is expanding, and if we assume that the rate of expansion was constant, the date of origin which is obtained is around 1140, not 1054. This discrepancy of 86 years is considerably larger than the errors due to the normal, and understandable, errors in measurement of the rate of expansion and the size of the nebula. A possible solution to these discrepancies is that there were two supernovae. But this explanation does not appear to work. There is no remnant southeast of Zeta Tauri which should be present if another supernova had occurred there. Nor is there a likely star which could have produced a nova, instead of a supernova (and, besides, the period of high brightness is too long for a nova). On the other hand, there are no records of a supernova near 1140.

The solution to the discrepancies in position and time of origin described above appear to have a straight—forward solution. In science, it is usually best to accept the simplest solution which does not have fatal problems. Firstly, the discrepancy in time of origin can be explained if the assumption of a constant rate of expansion is dropped. If the rate of expansion has accelerated, then the time of origin could come out correctly. For the rate of expansion to have accelerated, a source of energy is required in the nebula. As we shall see, there is a perfectly suitable source of energy in the nebula—the pulsar. Secondly, the Chinese record may be wrong, either by having the positions of the star (Zeta Tauri) and the supernova interchanged (an understandable error), or by simple mistake. Finally, when the Crab Nebula is assumed to be the remnant of the supernova of 1054, the astrophysical scenario works so well that it would be hard to accept that it is wrong.

RADIO AND X—RAY DISCOVERIES

Radio astronomy began with the first detection of a celestial radio source by Karl Jansky in 1931. The discovery was not followed up, however, because Jansky was an engineer working for the Bell Telephone Laboratories, and the discovery was an accidental result of his work. In the early 1940's, the American pioneer of radio astronomy, Grote Reber (1911—), detected an excess of radio emission coming from the general region of the constellation Taurus. He was unable to pinpoint its location in the sky, because of the poor angular resolution of his radio telescope.

By the late 1940's, however, superior angular resolution could be achieved by means of a technique called interferometry. An interferometer uses two radio telescopes at different locations operating in concert to observe the same object.   The Australian astronomers John Bolton and G. J.   Stanley, working in 1948, did not have two telescopes, so they used a trick: they used one telescope, located on a bluff above the ocean, to observe a source directly as it rises above the horizon and indirectly as its radiation is reflecting off the ocean up to their telescope.   Thus they simulated the action of a two—telescope interferometer.   In this way, they showed that intense radio waves were coming from the general direction of the Crab Nebula, and they dubbed this radio source Taurus A. Bolton and Stanley later were able to use a real two—telescope interferometer to refine the position and to strengthen the case for the identification of Taurus A with the Crab Nebula.   Though not the brightest radio source in the sky, it is the brightest supernova remnant in the radio region of the spectrum, and it was the very first radio source for which the optical counterpart was identified (except for the Sun, which is only "bright" in the radio region because of its closeness).

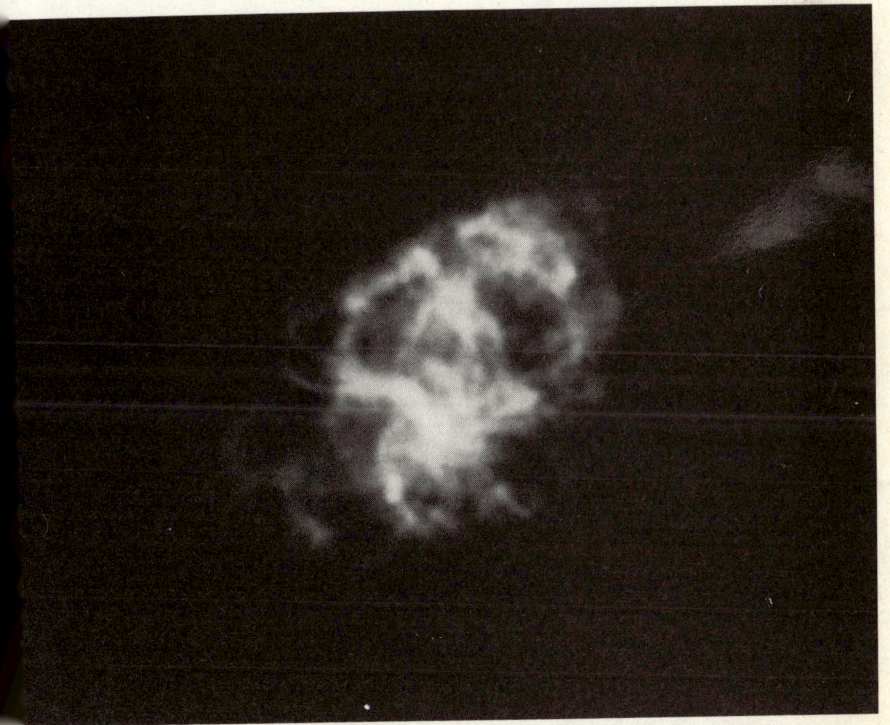

Figure 2.   The Crab Nebula, as seen at radio wavelengths.   The measurements are entered into a computer and displayed; the display is photographed to produce this picture (Courtesy NRAO/AUI).

In 1963 astronomers launched x—ray detecting devices on board rockets (which could parachute their instrument packages safely to ground).   It was necessary to send the detectors above the Earth's atmosphere, because the atmosphere absorbs x—rays, and does not allow any from celestial objects to reach the ground.   They detected strong radiation emanating from the constellation Taurus, and dubbed the source Taurus X—1.   When they were able to narrow down its location sufficiently, they concluded that, again, the Crab Nebula was the culprit. As in the case of radio sources, the Crab Nebula was the first x—ray source for which the optical counterpart was identified (except, again, for the Sun, the strongest x—ray source in the sky because of its closeness).

SYNCHROTRON RADIATION

The spectra of common gaseous nebulae are dominated by bright emission lines, characteristic of the fluorescence of a gas excited by ultraviolet radiation from a hot star within or near the nebula.   These same emission lines are present in the spectrum of the Crab Nebula, but they do not dominate the spectrum in the usual way.   The spectrum of the Crab Nebula is dominated by a continuum, as noted above.   This continuum is unusual, because it gets brighter as the wavelength gets shorter.   That is, the strength of the continuum emission grows from the radio region, through the infrared, the visible, the ultraviolet, and on to the x—ray region, where the Crab Nebula is the brightest object in the sky, (after the Sun).   The Crab Nebula was the first object known to show such a spectrum, and it was very puzzling at first.   An answer to this puzzle was proposed by the Soviet astrophysicist I. S. Shklovsky in 1953.   His proposed mechanism for the emission of the continuum is called synchrotron emission, and the emission occurs when electrons spiral at high velocities around magnetic field lines.   The name comes from devices called synchrotrons, which are used by physicists to study the structure of atomic nuclei.

Shklovsky's idea was not accepted until one of its predictions could be verified; that light emitted from synchrotron emission should be polarized.   It is easy to determine if the light from the Crab Nebula is polarized, and the observations were made by the Soviet astronomers V. M. Drombrovski, at the Byurakan Observatory in Armenia, and M. A. Vashakidze at the Ambastumani Observatory.   They verified that the light from the nebula is highly polarized; their observations were verified by W. Baade at the Palomar Observatory in 1955.

The remaining puzzle to be explained was the origin of the very energetic electrons.   Something was accelerating the electrons by a mechanism unknown in any other astronomical object. Some observations gave clues to the source of the energy. In the central region of the nebula are two stars, labelled NE and SW. As we have seen, the star labelled SW was at the center of the expansion of the nebula 900 years ago. Surrounding the two stars in the nebula is a complex of wisps and clouds. In 1921, C. O. Lampland, of Lowell Observatory, noticed that his photographs apparently showed changes in position and brightness of some of the wisps and clouds. These changes were studied further by Baade and Minkowski, and described in their paper in 1942. They were further studied by Baade until his death in 1960. Extensive work was also done by Scargle in the 1960's. The changes in brightness and position give the impression of "waves' moving outward from the center—the star labelled SW. It would thus seem that this star is the source of some kind of activity in the nebula. We will describe the source of this activity in Chapter 6.

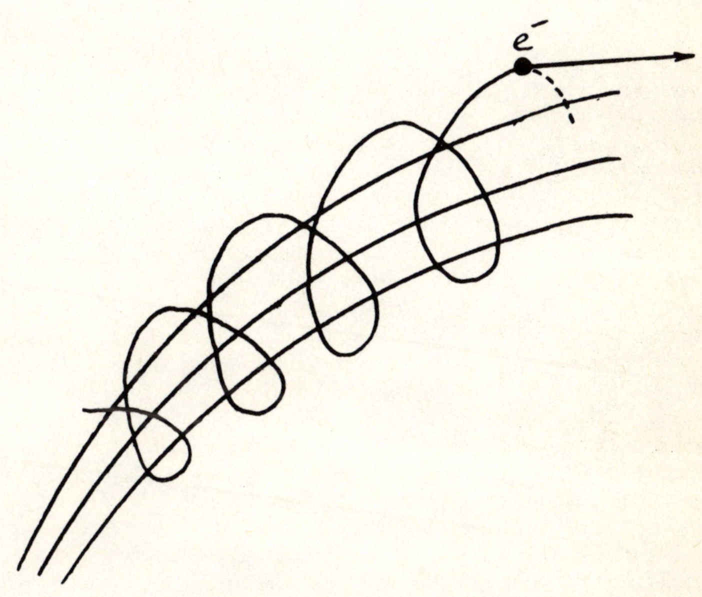

Figure 3. The mechanism of synchrotron radiation. The electron (e−) spirals around the magnetic field lines. As it moves, it emits electromagnetic radiation in the direction of its motion at every instant.

The remnants of other supernovae are much more difficult to study, so the Crab Nebula was the first to be studied intensively. Observations of the Crab Nebula established it as the remnant of the Supernova of 1054, and confirmed that a supernova is a titanic explosion in which a star explodes and blows off clouds of gas. The Crab Nebula is visible at wavelengths ranging from x—rays to radio waves, with characteristics which make it one of the most unusual objects in the sky. The Crab Nebula saga is an example, admittedly the best example, of how much astronomers can learn by studying the remains of an explosion which was observed long ago.

CHAPTER 5

SEARCHING FOR SUPERNOVAE

INTRODUCTION

Several new supernovae are discovered every year in other galaxies. They are usually quite faint, 10th to 15th magnitude or even fainter. While the highly detailed observations being made of SN 1987A are not possible, much useful information can be obtained from studies of these faint objects. Indeed, much of our knowledge of supernovae comes from discoveries in distant galaxies. Their light curves provide valuable information, and many of them are bright enough for some spectroscopy, although often with only modest resolution. Some of these supernovae are bright enough for the International Ultraviolet Explorer (IUE) to observe. The problem is to discover them to begin with so the large ground−based telescopes and the IUE can swing into action.

It might not seem difficult, at first thought, to find new supernovae, but in actuality it is a difficult observing challenge in astronomy. Many years can go by before a supernova appears in any given galaxy. Thus to have some chance of success, one needs to look at hundreds, or even thousands of galaxies with reasonable regularity. It is much more useful to catch a supernova while it is still on the rise, as opposed to finding it on the decline. Galaxies are not uniform objects. They contain bright nebulae surrounding hot young stars. There are foreground stars in our own galaxy that confuse things. Some of these foreground stars, to make things interesting, are variable stars. Also, asteroids can wander into the field of view; at opposition, asteroids move very slowly, and thus it may not be easy to recognize that they are asteroids.

Traditionally, supernovae have been discovered by placing a camera on a large telescope, taking a timed exposure of a field—preferably one rich in galaxies (a cluster of galaxies), developing the photograph, and then visually comparing it with a similar photograph taken at an earlier time.   While somewhat labor—intensive, and requiring considerable telescope time, the photographic method has, by far, been the most successful to date.

Recently, several supernovae, some still on the rise, have been discovered visually by amateur astronomers looking through very modest—sized telescopes. The Rev. Robert Evans, an Australian amateur, is the acknowledged champion in this specialized, but most useful area of astronomy.

For many years, searching for supernovae has been viewed as a natural task for automatic telescopes. For some time, Stirling Colgate and his associates were the only ones developing automatic telescopes for finding supernovae. More recently, several other teams have joined in this effort, although these new developments are for semiautomatic searches, as opposed to the more difficult fully—automatic task undertaken by Colgate.

VISUAL SEARCHES FOR SUPERNOVAE

As discussed in the third chapter of this book, several bright supernovae have been discovered visually. The Chinese astronomers were particularly good in making such discoveries, and more importantly, recording them for posterity. However, prior to 1987A the last naked—eye supernova was in 1604, and while it helped change European conceptions about the unchanging nature of the "fixed" stars, looking for naked—eye supernovae would not have been a very rewarding task between 1605 and 1986. While the telescope was invented just a few years after the 1604 supernova, and was used well by Fritz Zwicky (and others) to find supernovae photographically in recent years, the use of a telescope to aid the human eye in finding supernovae is very recent indeed.

The first supernova found visually with the aid of a telescope was 1968L (in NGC 5236). Jack Bennett (South Africa) found this supernova with a refractor having an aperture of only 0.12 meters in diameter. In 1979, Gus Johnson (United States) discovered another supernova. He used an 0.20—meter diameter reflector to discover supernova 1979C (in NGC 4321). However, it was Rev. Robert Evans who made visual discoveries of supernovae an approach to be taken seriously.

Evans made his first discovery in Australia in 1981 when he found the first supernova of the year, in NGC 1532 (1981A). He used a 0.25—meter aperture telescope to make this discovery. The Rev. Evans describes his telescope as a "Twenty—five centimeter portable Newtonian, f/4.3, of dubious and elementary construction." While his telescope is actually quite well suited to its task, it is his keen memory of many complex starfields, his charts and photographs to aid his memory, his dedication, and his persistence that are the keys to his phenomenal successes in recent years.   He has now discovered over a dozen supernovae—several of them on the rise.

Evans points out that the advantages of visual searches include low cost (volunteers), plenty of telescope time (small amateur telescopes), speed (no waiting for film development), and good coverage, including isolated galaxies that the photographic survey tends to skip over.

SEMI—AUTOMATIC SEARCHES

The basic strategy of semi—automatic supernova searches is to computerize the telescopes and data taking, but to manually examine the resulting data with the human eye and brain.   Those that advocate this approach suggest that telescope movements, sensor control, and recording of data are all processes conducive to automation.   On the other hand, comparisons of complex images for subtle differences is a process well suited to the human eye and mind.   In summary, let machines do what they do best, and let humans also do what they do best.

Those that advocate a fully automatic approach suggest that in terms of manpower requirements, human examination is too time consuming, resulting in (1) requiring more people for the project than are affordable; and (2) getting behind in the comparison process, resulting in discoveries being made later than they would be with a fully automatic process.

As both processes are now in operation, it should soon become evident which is most practical.   In the long run, as automatic telescopes, sensors, computers, and pattern—recognition software improve, it seems likely that the fully automatic approach will become dominant.   In the short run, only one team has a fully automatic approach in operation, and for the other teams a semi—automatic approach made practical sense in their circumstances.

THE CORRALITOS OBSERVATORY

The first working semiautomatic supernova search system was developed by the late Allen Hynek (Northwestern University), Justus Dunlap, and William T. Powers (New Mexico State University).   They built an observatory with a 0.6—meter telescope specifically to find supernovae.   Their observatory, known as the Corralitos Observatory, is located in New Mexico.

Their 0.6—meter telescope was computer controlled, and was able to slew to any one of many galaxies stored on a list in the computer.   Once a galaxy was acquired, an image orthicon detector was exposed to the light for the appropriate time, and the resulting image was recorded on tape for later human analysis.   While the computer slewed the telescope to the positions of the galaxies, an operator was still required for various aspects of the operation.

"Before" and "after" images were then subtracted electronically, and a human interpreter examined the results to decide whether or not something significantly new, such as a supernova, had entered the field.

The Corralitos system did work, and some 14 supernovae were found—the faintest being 19th magnitude. Sadly, funding was cut off, and the search was discontinued.

THE BERKLEY SEARCH

A group, primarily at the Lawrence Berkeley Lab (LBL), has been working for a number of years on an automated search system for supernova. While they originally intended to have it fully automated, they decided to scale back their goals and have achieved some success with a semiautomatic system. Team members include Luis Alvarez, Richard Muller, Carl Pennypacker, Frank Crawford, and Ronald Kahn, all of LBL, and Richard Treffers, an astronomer from the University of California (Berkeley).

The group took an existing 0.75—meter telescope and brought it under computer control. They added a Charge—Coupled—Device (CCD) camera to the telescope. Every clear night a computer directs the telescope to a series of galaxies. At each galaxy, an exposure is automatically made on the CCD camera, and the results are stored. This continues all night long. At some later time, a person views two digitized images. One was taken as a reference (no supernova), while the other was taken recently, and may or may not have a supernova (or some other new object such as an asteroid) in the field. The person looks at the two images and decides whether or not they are different.

The Berkeley system is currently in operation, and has discovered several new supernovae.

THE WHEATON COLLEGE SEARCH

One of the most interesting of the supernovae search programs is at Wheaton College, in Norton, Massachusetts. The brain child of Tim Barker, this project is the essence of simplicity (relatively speaking), and utilizes all off—the—shelf equipment—all modestly priced.

The telescope used in the project is a Celestron—14 Schmidt—Cassegrain on a DFM Engineering mount. An Astro Link intensified TV system is used as the detector.

The telescope and camera sit out in the cold, while a professor or student sits inside where it is warm. The computerized telescope slews to a galaxy, using pre—stored coordinates, and an image of the galaxy from the intensified TV camera appears on the video monitor. The operator then compares this with a transparency that is overlaid on the video monitor, and looks for any new stars.

The Wheaton system has the advantage that if a new star is found, identification will be essentially in real time, instead of from images recorded at some earlier time. It also has the advantage that the hardware and software are relatively simple. The main disadvantage of the current Wheaton College system is that the intensified TV system only reaches 14th magnitude at best, so the system misses the many supernovae that are fainter than this. Replacement of the TV camera with a cooled, integrating CCD camera should make a major improvement.

Figure 1.   The Wheaton College system for searching for supernovae. Devised by Tim Barker, this semiautomatic system uses low−cost, relatively simple equipment, and student observers. (Photo by Tim Barker, Wheaton College).

THE NEW SOUTH WALES SEARCH

A team at the University of New South Wales, headed up by Peter Mitchell, is developing a nicely−conceived automated search system. The telescope for the system is a converted Baker−Nunn camera. In the late 1950's, a dozen of these cameras were built by Perkin Elmer Corp. for the Smithsonian Astrophysical Observatory. They were placed at sites around the world, and were used to record the positions of artificial earth satellites. The Baker−Nunn cameras were not automatic, and were operated by crews for over a decade before more automated systems were developed by the Air Force.

The Smithsonian Institution donated the Baker–Nunn camera that was originally at the tracking station in Woomera, Australia, to the New South Wales group, and they have modified it extensively for their search program. They tilted the camera up to make it an equatorial mount, put on new bearings, and added motors and encoders to each axis.

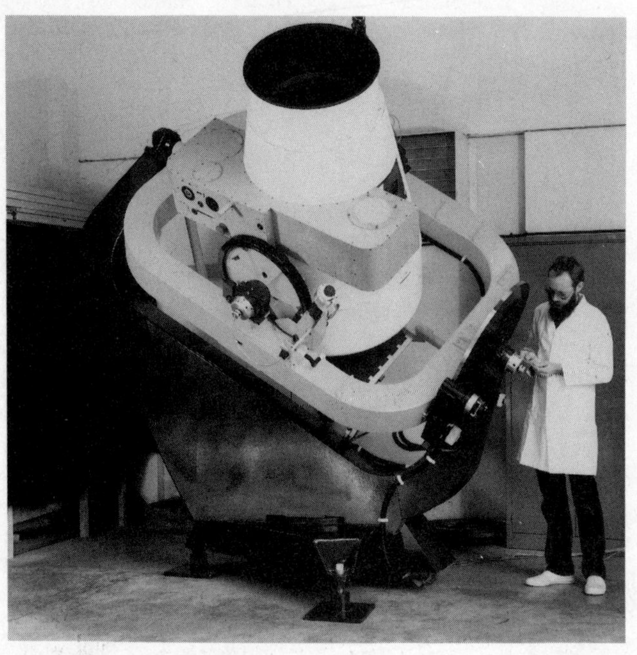

Figure 2.   The modified Baker–Nunn camera of the New South Wales group headed up by Peter Mitchell.   When completed, this should be the first southern automated supernova search system (Photo by University of New South Wales).

A cooled Charge–Coupled–Device (CCD) camera will record images of galaxies down to 19th magnitude, with fairly short exposures. The telescope and camera are automatically controlled by computers. While not yet operational, this system is progressing rapidly, and uses much current technology.

FULLY AUTOMATIC SEARCHES: THE COLGATE SYSTEM

For the past decade or longer, one could ask almost any astronomer at random the question "Automatic supernovae search?" and the answer, usually without hesitation, would be "Stirling Colgate." Colgate was responsible for one of the earliest successful theoretical models of supernova explosions. However, he also wanted to take an active part in getting the observational data needed to feed the theoretical models. In order to get observational data on supernovae, they first needed to be discovered—preferably early–on while they were still rising. Having used computers to good effect in modeling supernova explosions, Stirling saw no reason why they could not be used to find supernovae—all automatically and he took an active role in the design and construction of the equipment.   Thus the Digitized Astronomy Supernova Search (DASS) project was born.

In 1965 Stirling investigated the feasibility of several approaches to making an automated remote computer–controlled telescope at Socorro, where he was President of the New Mexico Institute of Mining and Technology.   He soon found the basic requirements of a digitally controlled telescope: a digital television readout, high speed digital transmission link, and sufficient digital computer capacity to deal with an ongoing supernova search; each in themselves represented a major investment in either money or technology beyond the then–known state– of–the–art.   The project was funded in 1967 by the National Science Foundation with the strong encouragement of Jerry Mulders and later Jim Wright.

The requirements of the system were set by the frequency with which Stirling felt he needed to observe distant galaxies to find supernova at a sufficient rate to be useful for their further scientific understanding. The human eye and brain is an extraordinarily effective system for pattern recognition and to hope to beat this with a computer–directed system would have to rely upon improved sensitivity and the ability to repeat a monotonous observation without fatigue. The basic requirement of this system was established by the frequency of supernova, the distance to galaxies, and the supernova brightness.   The conclusion was that several thousand galaxies had to be searched each night. Therefore the total time for image integration, image processing, and slewing would have to be less than ten seconds per galaxy. It was for these reasons that Stirling needed a 0.75–meter diameter telescope of low moment of inertia and very rapid slew rate. This telescope design requirement along with digital controls was beyond the then state–of–the–art.

Similarly the television sensors of the day, namely image orthicons, were barely adequate. The TV signal was an analog current and hence it was logically difficult to use in a digital system. The success that Hynek, Dunlap, and Powers had with image orthicons at the Corralitos Observatory was outstanding, and they found many supernova by using human observers to analyze analog television presented images. To do the same thing using a computer with the orthicon signal proved to be a complicated task. Colgate's group invested considerable effort, first in orthicons and then in isocons because they produce a positive signal and then finally (successfully) with an intensified silicon target vidicon, the ISIT—Vidicon.

The magnitude of these problems led Colgate and his group to the use of surplus military equipment, a tradition that had been developed at New Mexico Tech over many years for cloud physics in the laboratory established by Irving Langmuir. They planned on locating the DASS telescope at Langmuir Laboratory at 10,500 feet in the Magdalena mountains, twenty kilometers from the campus because of the dark skies and very good seeing. Stirling obtained a surplus Nike Ajax radar mount to use for his fast—slewing telescope. They decided upon a Schott—glass centrally—supported, lightweight mirror that was figured by Don Loomis of Tucson. This lightweight mirror was eminently successful.

The telescope was mounted on top of a steel tower that was a surplus Nike Ajax missile liquid oxygen tank. The radio link used for the high speed transmission of the data to and from the telescope used surplus as well as new equipment. The digital link had to operate at a very high data rate and again the commercial version by Bell telephone would have been prohibitively expensive.

The system hardware was completed in 1972 with the help of many students and technicians, John Colburn and Richard Carlson and the major design experience of Elliot Moore. Graduate students Jeff Colvin and Chester McKee obtained their degrees in theoretical astrophysics while associated with the project.

For a project like the space shuttle the software contracts were ten times the dollar value of the hardware. In retrospect, Stirling says this has been the case with his project. After the initial implementation of the hardware in the early 70s, the software and its interaction with the hardware became the central problem. The computer on the campus was an IBM 360 Model 44. They soon found that the architecture of a IBM 360—44 was not suited for real—time control because one needed overlays to store the immense programs that were involved. The difficulty in tracing errors, the failure rate of the telescope system hardware, as well as the IBM computer seemed almost insurmountable. At times in 1973 they had the whole system working, but consistent operation eluded them.

By 1974 the failure rate of the main computer contributed to their problems and at the end of 1974 Stirling left the Institute as its President after ten years. National Science Foundation funding was terminated in 1973, but, funds privately contributed as well as from the Institute allowed the software and hardware development and improvements to continue. In the fall of 1976, Stirling took a position at Los Alamos National Laboratory and with students at Tech, continued the development of the DASS system.

Figure 3.   Closeup of Stirling Colgate and the upper portion of his telescope.

Stirling decided that the only possibility for digitized astronomy was a time—shared virtual memory machine so that the multiple tasks to be performed by the telescope could be subdivided into multiple "users" in a multi—tasking environment.   This architecture of multiple, simultaneous tasks being performed such as slewing, readout, picture analysis, catalog selection and so on could be performed interleaved during the telescope operation.   For this purpose they selected the first minicomputer in this category, the Prime 300.   This computer was funded by the Naval Research Laboratory equally for atmospheric research and for automated astronomy.   Again they had to build an interface for this computer, but it was far less complex than for the IBM and so by 1977 they again established operations.   Fortunately, two very able undergraduate students (later graduate students), Eric Pearce and Kevin Meier, had joined the project.   They proceeded to develop the complicated software necessary for operating the system. This has been a long task and it has involved a complete rewrite of the software for digitized astronomy. The heart of the system is the program called "Direct." One of the most innovative software programs is the single pass processor of a digital frame "Find" for finding all stars in the field.

The DASS system is now operating and has been operating since February 1987. Over 15,000 galaxies have been searched, mostly in building the catalogs needed for comparison. The DASS system has found four test novae so Colgate's Group knows that the system will find supernovae when and if they grace them with their appearance.

CHAPTER 6

STARS AND GALAXIES

When the Moon is not up and the nearest city lights are many kilometers away, most people can see about 3500 stars in the half of the sky visible above the horizon at any one time. These same stars have been in existence for millions of years, many of them for thousands of millions of years. Each one of these points of light is really a ball of hot gas much like our Sun, which is also a star. The Sun is so bright simply because it is so close to Earth, only 150,000,000 kilometers away. To reach the next star, you'd have to travel 270,000 times farther away, a distance of 40 million million (40,000,000,000,000) kilometers. There you would find the small, faint dwarf star Proxima Centauri and, only a little farther away, the double star system Alpha Centauri. In this system two stars much like the Sun orbit around each other every 88 years, maintaining about the same distance from each other as the planet Uranus does from our Sun.

The stars, though apparently fixed in the sky, do move. They are moving in various directions through space with speeds of approximately 20 kilometers per second (neglecting their orbital motion around the center of the galaxy). Consequently, after several thousands of years, their arrangement in the sky, as viewed from Earth, will have changed somewhat and humans of later generations will not see the same familiar constellations as we do. In just 30,000 years, the very faint red dwarf variable star HH Andromedae (presently the 8th nearest star) will have replaced both Proxima and Alpha Centauri as our nearest neighbor. Because HH Andromedae is approaching us rapidly, it will later move away from us rapidly. Therefore, ironically, Proxima and Alpha Centauri (which move relatively slowly with respect to the Sun) will again be our nearest neighbors. This second switch will occur only 10,000 years or so later.

Figure 1. The spiral galaxy M31, in the constellation of Andromeda.

Figure 2.   The elliptical galaxy M87, in the constellation of Virgo.

In addition to their individual motions in various directions, the stars move together in a nearly circular flow about the center of our galaxy.   Our galaxy includes hundreds of thousands of millions of stars which orbit around their common center of gravity in a huge spiral—shaped pinwheel about 100,000 light years in diameter.   Our galaxy is an example of a type called a spiral galaxy.   Spiral galaxies are circular and flattened, and have a central bulge, rather like a fried egg (but bulging on both sides).   They are called "spirals," because when seen "face on," one sees spiral arms winding outwards from the central bulge (see Figure 1). In contrast, elliptical galaxies (see Figure 2) do not have a flattened disk, and do not have spiral arms; they are somewhat analogous to the central bulge part of spirals.   Irregular galaxies (see Figure 3) do not have a central bulge, and have no spiral arms.

The disk—shaped collection of stars, when seen on edge from a point away from the center, appears to us as the Milky Way, visible at night as a hazy irregular band of light in the sky stretching from horizon to horizon.   From any one location on the Earth's surface you can see only half of the Milky Way.   Our solar system (Sun, Earth, the other eight planets, meteoroids, comets, *etc.*) is moving around our galaxy's center, at a speed of 220 kilometers per second, in a nearly circular orbit. This orbit is similar to the Earth's orbit around the Sun.   Just as the Earth takes one year to complete an orbit around the Sun, the solar system takes 250 million years to complete an orbit around the center of our galaxy.

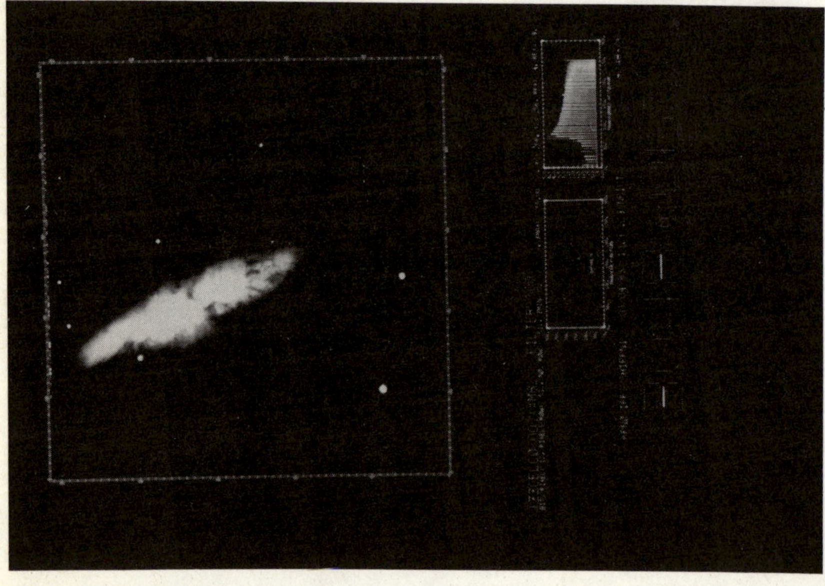

Figure 3.   The irregular galaxy M82.

Supernova 1987A occurred not in our galaxy, but in the Large Magellanic Cloud, a rather small galaxy nearby our own. It and the even smaller Small Magellanic Cloud, are our nearest neighboring galaxies, and are both visible to the naked eye from the southern hemisphere. Both the Large and Small Magellanic Clouds are irregular galaxies, not spirals as ours is, and they contain only about 10% as many stars as our own. They were named after Ferdinand Magellan, the Portuguese explorer who led the first expedition to circumnavigate the Earth in the early sixteenth century. That voyage carried him and his crew far enough south to let them see those two galaxies clearly in the southern sky. They can't be seen from locations on the Earth's surface farther north than a latitude of 20 degrees, where they are right on the southern horizon. Thus, one must be located south of Mexico City, Hawaii, Calcutta, or Saudi Arabia to see them at all.

Although there are thousands of millions of galaxies in the observable universe, most are known to occur in groupings of various sizes, called simply "clusters of galaxies." Our galaxy belongs to a relatively modest cluster called the "Local Group." Its membership of 24 includes both the Large and Small Magellanic Clouds and the three full—sized galaxies: M31, M33, and our own.

The distances between stars, compared to their sizes, are immense. If stars like our Sun were grains of sand, they would be separated from each other in all directions by distances of about 30 kilometers. Galaxies, on the other hand, are much, much closer together in proportion to their diameters. Galaxies in the Local Group are separated from each other by distances no larger than 10 or 20 galaxy diameters and as little as a few; some are almost touching each other. Curiously, metropolitan areas in the Eastern United States are separated from each other in roughly the same proportion. Small satellite galaxies like the Large and Small Clouds could be compared to smaller towns near large cities. Just as urban sprawl often includes metropolitan areas touching each other, some galaxies are probably joined by a bridge of stars. Actually, just such a star bridge is thought to arch between our Galaxy and the Large Magellanic Cloud.

Supernova 1987A in the Large Magellanic Cloud is about 170,000 light years from the Earth. Although in a different galaxy, it is not all that much farther away than a supernova in our galaxy might be. The center of our galaxy is about 35,000 light years away and a star on the far side of our galaxy would be about 85,000 light years away. Supernova 1987A has an additional advantage that, because of its favorable orientation with respect to us and our flat galaxy, relatively little of our galaxy's interstellar dust lies between it and us. Thus the total extinction of visible light along the 170,000 light—year—path amounts to about 60%. This amount of extinction is about three times that which the light from a star suffers during the last millisecond of its long journey to Earth, through the Earth's own atmosphere. That explains why this star in another galaxy was able to become one of the 200 brightest stars in the entire sky, even though at least one recent supernova in our own galaxy was not seen at all.

MODELS OF STARS

What are stars? Are new stars being born now, or were they all created at some time in the past? Are they always the same, or do they change as time passes? Why do some of them "blow up" in supernova explosions? The answers to these questions are pretty well known, now, but finding them has not been easy. Although we have answers to these questions, they are not final, and they raise more new questions. The study of stars as objects to be understood physically started about one hundred and fifty years ago. Before then, stars were only points of light in the sky. They could be described with their positions, motions, and brightnesses. As physical objects they obeyed Newton's Laws, but any questions about their physical nature could only be answered with speculation. Observational data and the laws of physics necessary for an answer were not then known.

The nearest star is the Sun, and it is a typical one in many ways. The nearest star beyond the Sun is so far away that it is seen as a mere point of light, even in the largest of telescopes. Its apparent brightness is a million million times smaller than the Sun's. Thus it is not surprising that the physical study of stars started with the study of the Sun. Its great brightness makes physical study much easier, but even in the case of the Sun, it is not easy. For essentially all of our information we must rely upon the light (and other electromagnetic radiation, such as radio waves, infrared radiation, ultraviolet light, *etc.*) we receive from it. No spacecraft has visited the "surface" of the Sun; and even if one had tried, the apparent surface is at such a high temperature that the spacecraft would have been vaporized before reaching it.

Our knowledge of the interior of the Sun must depend primarily upon what we can learn from the light we receive from its "surface." The method of learning about the interior of the Sun and stars from the characteristics of the light coming from their surfaces depends on a technique called modeling. It depends upon calculating a model of the interior using laws of physics discovered in laboratories here on the Earth or derived theoretically from laboratory data. The laws of physics which are thought to apply to the interior of a star are used to calculate the conditions at many points from the center out to the surface of the star, so that we can make a prediction of what the surface of the model star will look like from a distance. This prediction is compared with observations of the surfaces of real stars. One almost always finds that the predictions fail to some degree; usually the way the prediction fails gives us a clue about what is wrong, or what is incomplete about the model. The model is then revised, and a new comparison is made.

Attempts to calculate model stars began roughly one hundred years ago, but the first attempts were very primitive. At that time neither the physical laws nor the measurements of stars needed to calculate correct models were known. Since then, the study of atomic and nuclear physics has provided many of the physical laws that were needed, and the astronomical observations have provided the necessary knowledge of stars. The result is that we now have a pretty good idea of not only what the interior of the Sun is like, but of how it has changed as the Sun has evolved over time, starting soon after its "birth" about five thousand million years ago, to its "death" somewhat over 10 thousand—million years in the future. This achievement, the ability to calculate the interior structure and evolution of stars, must be counted as one of the major intellectual achievements of the twentieth century.

In this chapter, some of the results obtained in this achievement will be described. We will start out by describing the basic nature of the Sun, a typical star. After that we will talk about other stars.

THE SUN

As we look at the Sun, we recognize that it is an intense source of light and heat. Even at the distance of 150 million kilometers it dominates the daytime sky and keeps our planet warm enough for life. A calculation of the temperature at the visible "surface" gives an answer of about 6,000 centigrade—high enough that every element is a gas. In fact, we know that the Sun is entirely gaseous. Why, then, does it appear to have a "surface," or a sharp "edge"? Why can't we see through it? You might ask this question, because the Earth's atmosphere, the largest body of gas we are acquainted with in daily life, appears to be transparent, at least when clouds are not present. The fact is, though, that even the Earth's atmosphere is not completely transparent. You know that distant mountains look "hazy" and "purplish," this results both from the lack of transparency of the gases in the atmosphere and from the scattering of light from dust particles in the air. If you could look through a long enough path (which the curvature of the earth makes difficult) you would find that some of the light does not get through. We can't see through the Sun because the gas of which it is constituted is also not entirely transparent. You might, then, expect the Sun to have a "fuzzy" edge, rather than appearing to have a sharp one. This results from the fact that as we see the Sun in the sky, the region over which it would appear to be "fuzzy" is such a small fraction of its total diameter that it appears to have a sharp edge.

It will help if we get an appreciation of the vast size of the Sun. Let's compare it with the Earth. The Earth, itself, is enormous compared with the regions of our daily experience; the diameter of the Earth being about 13,000 kilometers, and the circumference about 40,000 kilometers. The Sun, also a spherical globe, has a diameter about 110 times as large.

The amount of material in the Sun, which we call its mass, is about 330,000 times the amount of material in the Earth. It may be hard to visualize something gaseous being so much more massive than a ball of rock and iron, such as the Earth. The great mass of the Sun can be thought of in terms of its vast size plus the fact that the gases in the center are greatly compressed by the force of gravity to tremendous densities. The Sun appears to be of modest size, as seen in the sky, only because of its great distance from us.

What holds this great ball of gas together? The same force which holds the atmosphere, and us, for that matter, on the Earth—gravity. The theoretical proposal that the Sun is a great sphere of gas held together by gravity, and that the gravitational force was balanced by the pressure of the hot gas, was first published by the English physicist J. H. Lane in 1870. As a result of its great size the weight of the outer layers compress the inner layers to tremendous densities, far above anything we can generate on the Earth. How can it be a gas, then? The high densities are the result of the great pressure in the interior layers of the Sun, and if the gas were cold, the great pressure would compress them into a liquid, or, even into a solid. This does not happen because the temperatures in the interior regions of the Sun are also very high—about 20 million degrees centigrade at the center. At these great temperatures, all elements are gases, even at great pressures and high densities. The realization that the material deep in the interiors of normal stars was a gas, in spite of the enormous pressures, was due to Sir Arthur Stanley Eddington (1882–1944), of Cambridge University in England. His work produced the first successful models of the interior of the Sun. Eddington was a theoretician, which means that he concentrated upon understanding astronomical observations by making mathematical calculations using physical laws, rather than by making observations.

One of the most obvious and valuable characteristics of the Sun was one of the most difficult to explain, initially. The Sun radiates light and heat—what is the source of all of this energy? Several possibilities, known to physicists and astronomers at the turn of the century, such as chemical burning, were completely inadequate. On the basis of geological observations of the Earth, and assuming the Earth and Sun to be the same age, the Sun was thought to be very old. We now accept an even longer age for the Earth and Sun, about five thousand—million years, but the age deduced then was considered to be incredibly long. If the Sun's energy came from the burning of gasoline, it would "run out of gas" too quickly. With the current value of the age of the Sun, chemical burning is completely inadequate. Other explanations, such as radioactivity and gravitational contraction were tried, and none worked. The answer, nuclear reactions, was suggested in the 1920's, but was not well understood until after 1939 (see Chapter 8). Nevertheless, great progress was made in understanding the nature of the Sun and stars even before the source of energy was understood. This points out one of the remarkable things about a science such as astronomy—it is not necessary to have all the answers to a problem to make progress in its understanding.

CHAPTER 7

THE NATURE OF THE STARS

WHAT STARS ARE MADE OF

Figure 1.   The spectra of Vega.

An important characteristic of a star is its composition—how much of each element is present in the star. The composition of the stars cannot be determined directly, because there is no way of taking a sample of the material in a star. The key to determining the composition of the Sun is found in the observational technique called spectroscopy. The spectrum of the Sun or a star is crossed by dark lines called absorption lines. In the case of the Sun, many thousands of such lines are seen, some very strong, and many very weak; this is true of other stars such as Vega, whose spectrum is seen in Figure 1. One of the great astronomical discoveries of the nineteenth century was made by the German scientists Joseph Fraunhofer (1787–1826) and Gustav Kirchoff (1824–1887). Fraunhofer studied the dark lines in the spectrum of the Sun and showed that they occur in the spectrum of the stars and some planets, and Kirchoff found in 1859 that some of the lines formed patterns which matched those of known elements produced in laboratories on Earth. This act of recognition initiated the technique of chemical analysis by spectroscopy. In the decades that followed Kirchoff's discovery, many of the elements known on the Earth were recognized to be present in the atmosphere of the Sun. One element, Helium, however, was first discovered in the Sun spectroscopically, and only later was it discovered on the Earth.

The chemical analysis initiated by Kirchoff was a great advance, but it suffered the limitation that it revealed only what was present, not how much. The ability to deduce the relative quantities of the elements depended upon the advances in atomic physics which occurred during the early decades of the twentieth century. The science of quantitative stellar spectrum analysis became mature with the work of the Indian astronomer Megh Nad Saha (1893–1956), the English–American astronomer Cecilia Payne (later Payne–Gaposchkin; 1900–1979), who did her work at Harvard University, and Henry Norris Russell (1877–1957), who was a professor of astronomy at Princeton University. This work culminated in the recognition by Russell, in 1929, that the element hydrogen was the most common element in the atmospheres of the Sun and stars. Hydrogen is the simplest of the atoms, consisting of a single electron circling a single proton. By contrast, oxygen consists of 8 electrons circling a nucleus containing 8 protons and 8 neutrons. Hydrogen is a gas at the temperatures found on the Earth, and is extremely rare in its elemental form. It is a relatively common element on the earth, however, because it forms molecules with other elements, and is a constituent of the water molecule, H_2O. Still, as common as water is on the surface of the Earth, hydrogen is only a minor constituent of the Earth as a whole.

Astronomers find that the common stars in the neighborhood of the Sun have formed out of gas containing about 72% hydrogen (by mass), 25% helium, and about 3% heavier elements. The heavier elements include those which are most important to us on the surface of the Earth. Our atmosphere is mostly nitrogen and oxygen. Water is mostly oxygen by mass. Rocks include silicon, oxygen, and many metal atoms, such as aluminum, iron, copper, and so forth. It is thought by geophysicists that the core of the Earth is made primarily of a liquid consisting mostly of iron with a certain amount of nickel mixed in.

So, taken as a whole, the composition of the Sun is very different from that of the Earth. It is found that the other stars are generally very similar to the Sun, at least in their outer layers.

Once the composition of the stars was understood, it became possible to apply the theoretical ideas of Eddington to construct models of their internal structure. The source of the energy radiated by the stars was still unknown, although by the time (1929) the composition of the stars was understood, many astronomers and physicists believed that the source of energy involved nuclear reactions.   The fact that nuclear reactions could release energy is based upon the theory of relativity published by the German—Swiss physicist Albert Einstein (1879—1955) in 1905.   In that theory, he derived the famous equation $E=mc^2$, which says that mass and energy are equivalent and interchangeable.   In the case of the nuclear reactions occurring in the Sun, a series of reactions occurs in which four protons (the nuclei of four hydrogen atoms) are combined to form a nucleus of a helium atom (two protons and two neutrons).   The mass of the helium nucleus is slightly less than the mass of the four protons which were used to produce it.   What happened to the extra mass?   It was released as energy.

What is remarkable about this transformation is that a little bit of mass will produce an incredible amount of energy.   If you could transform a single penny entirely into energy in the form of electricity (not currently possible), you could supply your house for about 5000 years.   This amount of energy would be worth several million dollars, at current rates. The amount of energy being radiated from the surface of the Sun every second is beyond comprehension on the scale of our daily life.   If converted directly into electricity and sold at current commercial rates, it would take the entire budget of the U. S. Government for a thousand million years to pay for it.   This is just the energy radiated in one second!   All of this energy is generated by nuclear reactions occurring in the central regions of the Sun.   Further, these reactions produce energy so efficiently and there is so much fuel that there has been no difficulty in continuing for the five—thousand million—year lifetime (so far) of the Sun. The fact that this could be true was recognized before the details of the process were discovered; the detailed process was first worked out by the German—American physicist Hans Albrecht Bethe (1906—) in 1939.

KINDS OF STARS

The Sun is a fairly typical star, for its neighborhood in our galaxy although it is brighter than most.   The stars we find near the Sun are similar in their original chemical composition, but they do exhibit a great variety in other characteristics.   In particular, they differ from each other in two important ways: their masses, and their evolutionary state.   We will discuss their masses, their brightnesses and colors, their spectra, and briefly introduce their evolution.

The mass of a star is a measure of the total amount of material in the star.   In the case of the Sun, we determine the mass from the characteristics of the orbits of the planets about the Sun.   That is, we use the gravitational attraction, which affects the relation of the period and the size of the orbits of the planets, to determine the mass of the Sun.   We have not yet found similar planets orbiting other stars, so this method cannot be applied to other stars.   But the same principle can be applied in cases where there are two stars orbiting each other.   Thus, binary star systems can provide information on stellar masses. Fortunately, binary stars are very common, and it is possible to find some cases in which all the right observations can be made so as to get the masses to high accuracy. The observation of binary stars, and the theoretical description of their orbits and the methods of extracting masses, was another field of contribution by Henry Norris Russell early in this century.

We have since accumulated enough data to allow some statistical conclusions to be made; we find that stars apparently are limited to masses from about 1/20 of the Sun's mass to a few hundred times the Sun's mass. An interesting and significant relationship, which is found for stars in the neighborhood of the Sun, is that the smaller the mass of the stars, the more common they are. Another way of saying this is that very massive stars are rare. However, when we look out at the stars at night, many of the stars we see are very massive. This is because they are also very bright, so we can see them from greater distances than we do stars of small mass.

When we look at the sky at night, we see stars with a variety of brightnesses and colors. The apparent brightness of a star is the result of two main factors, its intrinsic brightness, and its distance from us. Its intrinsic brightness is related to the amount of light radiated by the star from its surface every second. We can compare the intrinsic brightnesses of different stars by comparing how bright they would appear if they were located at the same, standard, distance from us. If we can determine the distance of a star, then with its apparent brightness we can determine its intrinsic brightness. The intrinsic brightness is generally measured in terms of absolute magnitude. The colors of stars, which cover the range from red through violet, are evidence of another intrinsic characteristic of the stars: the temperatures of their surfaces. The surface temperatures of the common stars range from about 2500 degrees centigrade for the very red, cool ones like Betelgeuse to about 25,000 degrees centigrade for the most extreme bluish—white, hot ones like those in the Trapezium in Orion. The Sun is a yellow—white star with a surface temperature of about 6000 degrees centigrade.

The surface temperature of a star is also related to another observed characteristic of stars: their spectral type. Around the turn of this century, before the physical processes which determine the spectrum of stars were understood, stellar spectra were classified into a rather arbitrary sequence to which the letters of the alphabet were assigned. After the relationship with surface temperature became understood, the sequence came out to be OBAFGKM (in the order: hot to cool). O—Type stars are very hot: about 25,000 degrees centigrade. The Sun is a G—type star. Betelgeuse is an M—type star. This sequence of spectral type has been further broken down into subtypes. For example, G—type stars can be G0, G1, G2, ... G9, but not all of the possible subtypes are used. The Sun is a G2 star. When the intrinsic brightness and the colors of stars are plotted against each other on a graph, correlations are found which tell us much about the interior structure and evolution of the stars.

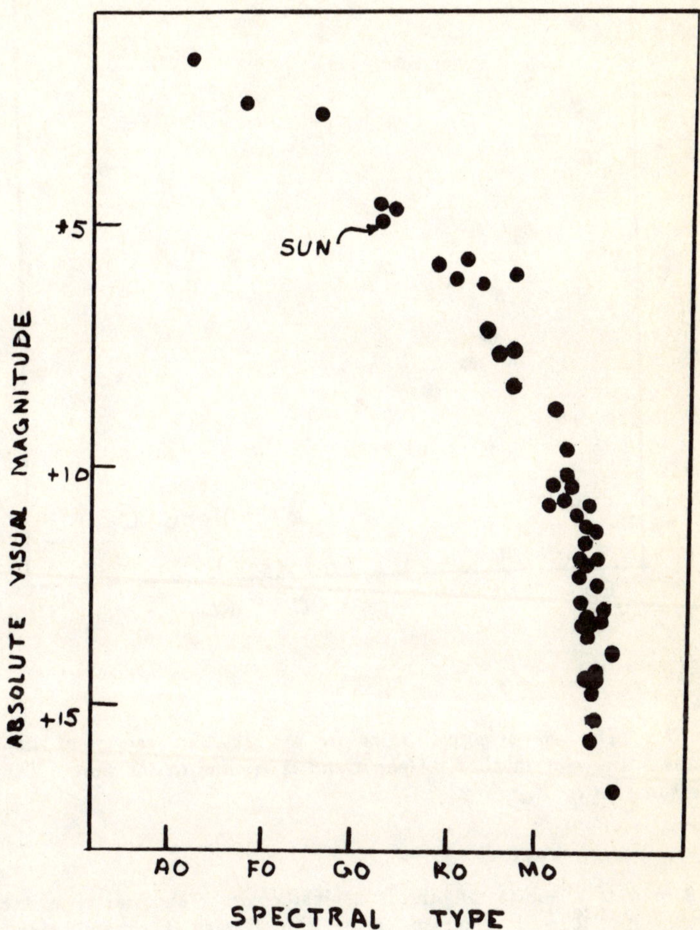

Figure 2.   An HR diagram of the nearest stars.   All stars within a distance of 150 million million kilometers are plotted.   The absolute magnitude (related to intrinsic brightness) is plotted against spectral type. The location of the Sun is indicated.

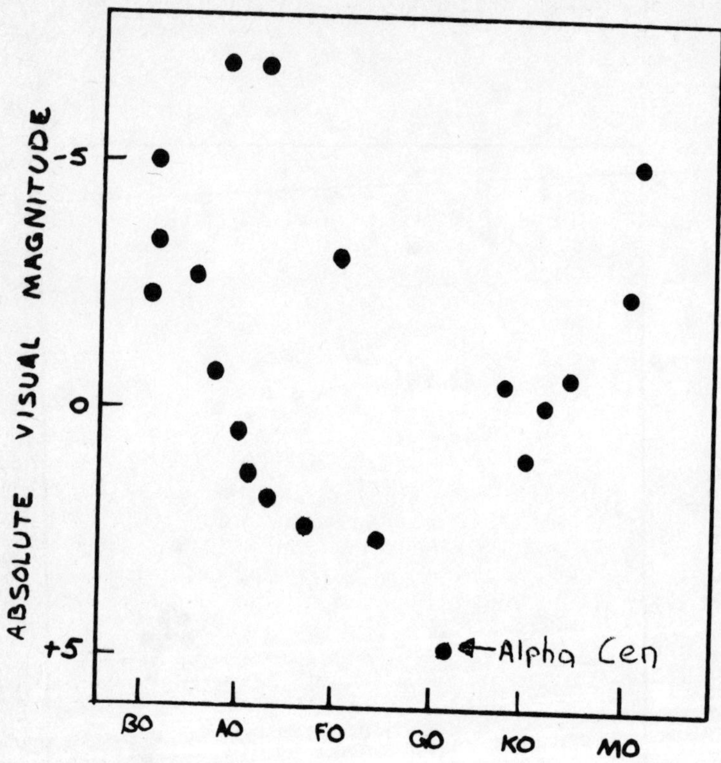

Figure 3.   The 20 brightest stars in the sky, in terms of apparent brightness.   The star labelled "Alpha Cen" is similar to the Sun.   The axes are the same as in Figure 2.

A plot of absolute magnitude against color or spectral type is called a Hertzsprung—Russell diagram, after the Danish—German astronomer Ejnar Hertzsprung (1873–1967) and Henry Norris Russell.   In the years soon after 1910 they independently found the same significant characteristics of the distribution of stars in this diagram.   Russell's publications of 1913 received the greatest notice.   An HR diagram, as it is commonly called, is shown in Figures 2 and 3.   Three characteristics of this diagram should be noted.   One is that the majority of the stars fall in a band lying diagonally from the upper left (blue, bright stars) to the lower left (red, faint stars).   This band is called the main sequence.   It is found to be the region in which stars are deriving their energy from the conversion of hydrogen into helium, as described earlier. Second is a group of stars in the upper right hand region of Figure 3, (red, bright stars). As noted above, we have found that the color of a star correlates with the temperature at the "surface" of the star, so that the bluer stars have hotter surfaces than the yellow stars (such as the Sun), and these have, in turn, hotter surfaces than the red stars.

The intrinsic brightness of a star is a result of two factors: the surface temperature of the star and its surface area. Remembering that surface area of a sphere is proportional to the square of the radius, we are saying that of two stars with the same surface temperature, the larger one will be the brighter. This conclusion can be applied to the stars in the upper right hand corner; they have the same color as main—sequence stars below but are brighter, therefore they must be larger. The main sequence stars are commonly called dwarfs; those in the upper right are called giants. The reddest are called red dwarfs and red giants, respectively. The stars shown here include relatively nearby stars, if we included a wider sample of stars, we would find some even brighter; these are called supergiants. We would also find that the giants and supergiants include some bluer stars, but that these are much rarer than the redder kinds. Finally, we see two stars in the lower left—hand corner of Figure 2. These are bluer stars, and therefore hot, and yet they are intrinsically very faint. They must have radii much smaller than the main—sequence stars. They are called white dwarfs, and we will discuss them later.

The vast majority of the stars are remarkably constant in their brightness, our Sun being one such example. A small fraction of them, however, vary in their brightness due to one of a variety of different mechanisms. Astronomers call these variable stars. In principle all stars are variable, because it is difficult to imagine any light source, natural or man—made, that could be perfectly steady. Probably about 5% of all stars, however, vary in brightness by an appreciable amount, let us say by more than about 10%. It is certain that the first variable stars ever discovered and recorded, many centuries before the birth of Christ, were novae or supernovae, but it is curious that these dramatic stars were not immediately categorized as variable stars. The first variable star recognized as such was Omicron Ceti, a bright star in the constellation Cetus (the whale). The discovery was made by David Goldsmith, an English astronomer who used the Latin pen name Fabricius. Fabricius noted that Omicron Ceti varied repeatedly from 2nd magnitude to 9th magnitude and back to 2nd magnitude approximately every eleven months. Another name for Omicron Ceti is Mira, which in Latin means "the wonderful one" or "the miraculous." Although the French astronomer Emilio Bouliaux in 1667 proposed the theory that Mira varies in brightness as a result of starspots (similar to sunspots which darken the surface of our Sun, but in this case more severely), his theory was ultimately (centuries later) proven wrong. Astronomers now know that Mira, and other stars like it, dubbed Mira—type variables, are pulsating red giants.

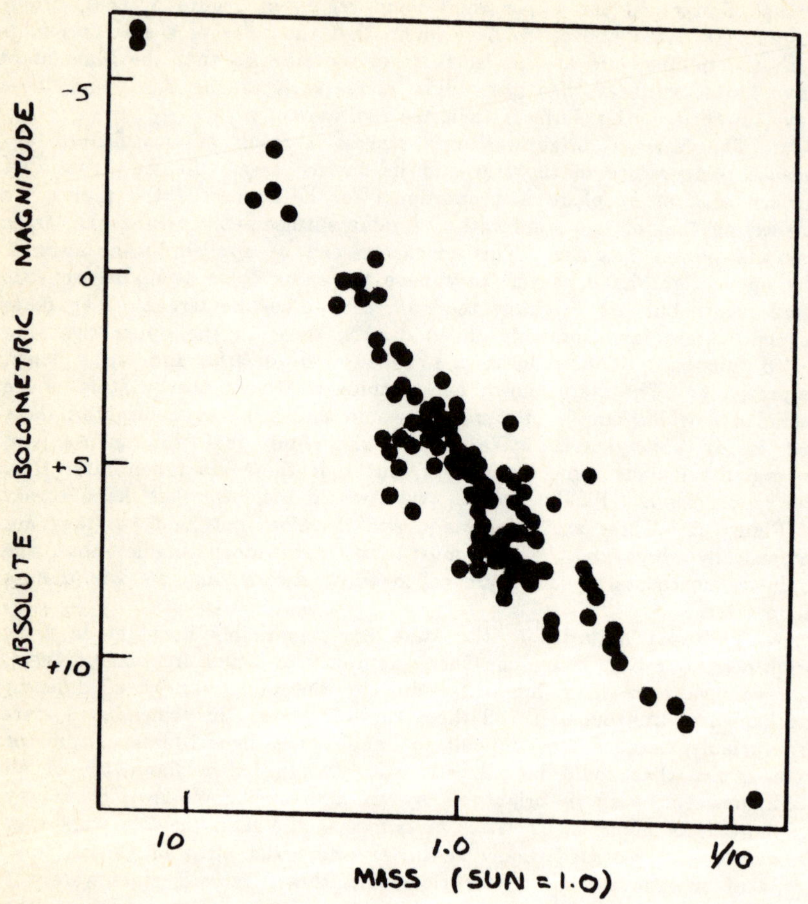

Figure 4.   The mass—luminosity relation for main—sequence stars.   The absolute bolometric magnitude (the luminosity expressed in units of magnitude) is plotted against the logarithm of the mass.

The *intrinsic brightness* of a star is usually referred to as the brightness it would have if seen by the human eye at a certain standard distance.   The human eye is sensitive to a certain range of wavelengths, but not to the entire range radiated by a star.   That is, a star will radiate heat energy, called infrared light; it will also radiate ultraviolet light (sometimes referred to as "black light").   Neither infrared nor ultraviolet light is visible; we feel the former as heat radiation, and feel the effects of the latter as the rays which produce sunburn. A cool star can radiate much of its energy as infrared light, and a hot star can radiate much of its energy as ultraviolet light, so the "intrinsic brightness" referred to above will be an inadequate measure of the total energy output of many stars. The total energy output is called the star's luminosity. It is equal to the total amount of energy produced by the nuclear reactions occurring in the interior of the star, as long as the star is not expanding or contracting.

When the luminosities of the stars are plotted against their masses, another correlation appears. The resulting diagram is called a mass–luminosity diagram, and it was first presented in its modern form by Sir Arthur Eddington in 1924. He pointed out that the luminosity and mass were closely related for all of the stars which he considered. He also presented a theoretical explanation of this behavior; it involved the conclusion that the stars behaved as if the material in their interiors was entirely gaseous. The correlation which Eddington presented says that the luminosity is roughly proportional to the cube of the mass of the star. This result is remarkable, and has a consequence for stellar evolution which is of the greatest importance.

When we talk about the evolution of a star, we are referring to how it changes over time. The main cause of a change in a star is the consumption of hydrogen in its interior by the nuclear reactions converting hydrogen into helium. The structure of the star changes as the interior regions become more completely dominated by helium instead of hydrogen. Finally, a point is reached at which the hydrogen is exhausted in the regions in which the reactions were occurring, and this makes the generation of energy much more difficult. When this occurs, the star must change its structure radically. Thus, the amount of nuclear fuel supplying the "furnace" in the interior determines its future. The amount of fuel, which is hydrogen in the case of main sequence stars, is proportional to the mass of the star. That is, the more massive a star is, the more fuel it has available to burn. You might think that this means that it would last longer, but this is not true. It turns out that the more massive stars use their fuel faster, much faster. The rate at which fuel is used is proportional to the luminosity of the star, which we just found is proportional to the cube of the mass. Thus, if we compare a star of two solar masses with the Sun, we find it has twice as much fuel, but uses it up at eight (two cubed) times the rate. This means that it will keep going only one–fourth as long! This is a general conclusion: the more massive the star, the shorter its "nuclear–burning" lifetime. Another way of saying this is that more massive stars evolve faster than less massive stars.

CHAPTER 8

THE LIFE CYCLE OF THE STARS

REAL STARS AND MODEL STARS

At the beginning of the last chapter, we posed the question: are new stars being born now, or were they all created at some time in the past? Astronomers are convinced that stars are being born now, and that the creation of new stars is a normal, continuous activity in our galaxy. This process has been going on for at least 10 thousand—million years (more than twice as long as the Sun has existed), and will continue almost indefinitely. Thus, around us we see stars of all ages—very young and very old, and everything in between. How are we going to make sense of what we see? As we see stars in the sky, they all look like points of light—what distinguishes a young one from an old one? It turns out that there are some distinguishing features, but they are generally quite subtle. Piecing together the story of the birth, evolution and death of stars has been challenging, and we will discuss a few of the steps in the process.

In the following sections, the life history of a star will be detailed. At each stage in the evolution of the star the physical processes going on in the interior will be described. But here, as in any astronomy book you read, you must realize that what is being described is not what a real star does, but what a model star does. We have never seen a star "do" most of these things, because the time involved is many human lifetimes. The Sun, for example, will pass through a number of significant and distinct stages during its evolution, but each stage lasts typically from millions to thousands of millions of years. So, during a human lifetime it will look the same for year after year. Some stars do occasionally do something quickly, and one characteristic of objects as large as stars is that if something happens quickly, it also happens violently—such as a supernova explosion. But, you needn't worry; the

Sun will not blow up—it is not the kind of star to do anything so spectacular. How do we know this? After all, it is pretty important to be right! Actually, we don't, but we do know what model stars with characteristics like the Sun's will do. And the model stars now match the Sun, or stars thought to be like the Sun, well enough that we are pretty confident that we know how the Sun will behave. This does not mean that we have any final or complete answers, even about the Sun. There are some important ways in which the models we calculate for the Sun disagree with the observations. Always, in such cases, there is the question of whether the problem is in the model or in our carrying out the interpretation of the observations. Even with these questions unanswered, though, we now know much about the evolution of the Sun and many similar stars.

As we consider stars with characteristics very different from the Sun's, our knowledge becomes much less certain. Examples are stars with extremely large masses, or stars which have evolved through many stages of their evolutionary life cycle, or stars undergoing some violent event, such as a supernova explosion. Most of the events and processes described in the narrative of stellar evolution given below are pretty well understood. When we get to the description of the supernova explosion itself, though, we must remember that what is being described is the best guess we have, and the results of further observations and the construction of new models might give different answers, at least in part. It is especially true that observations of a bright supernova, such as 1987A, have given us information which have changed our models in important ways. Another aspect of stellar evolution about which we do not know all the answers is at just the opposite end of the cycle from the supernova explosion: the birth of stars. We know enough, and can compare with observations in a sufficiently definite way, to know the general process of star formation, and that will be described below. But there are many important details, and a few significant questions, about which we are still mystified.

THE BIRTH OF STARS

Imagine a great cloud of gas and dust in interstellar space. In appearance, it is not too different from clouds in the Earth's atmosphere, except that it is dark—not lit up by the Sun—and not made up of water droplets in air. The composition is important—about the same proportions of hydrogen, helium and heavier elements given in the last chapter. The hydrogen and helium are gases, in spite of the extreme cold of interstellar space. But the heavier elements are not all in the form of gases. Some are locked up in tiny "dust grains." Interstellar dust is made up of many small particles consisting of a core of stuff like dust and soot on the Earth. That is, much of the dust on the Earth is ground—up rocks, made of what are called silicates. The silicates in interstellar dust grains probably have a somewhat different composition and different structure than dust on the Earth.   Soot, the other constituent, is carbon, packed together loosely.   The carbon in interstellar dust has a different structure than soot on Earth, because it is thought to be mostly a form of graphite—the stuff of pencil lead.

Around the core of silicate or graphite, the typical interstellar grain has a mantle of ice (mostly water ice, although other materials, such as methane and ammonia can be ices at these temperatures).   Also present will be more complex molecules, which include many which are considered "organic" molecules.   All of this material is present in particles too small to see with the unaided eye as individual grains.   We know the grains are present in interstellar space, however, because they scatter starlight, making a celestial "haze."   This interstellar extinction, as it is called, is a nuisance for astronomers trying to observe distant stars.   So also is another effect, called interstellar reddening, which is just another aspect of the interstellar extinction.   For example, we are already very familiar with a similar effect because on the Earth, when the Sun is rising or setting, it looks redder and dimmer than when it is more nearly overhead.   This is because the molecules and dust particles in the Earth's atmosphere scatter blue light more than red light, so relatively more of the red light gets through.   A similar process occurs in interstellar space, so that stars seen through interstellar dust clouds appear redder than they are intrinsically.

A cloud in space is in continual motion, just like a cloud in the Earth's atmosphere.   The motion is turbulent—a slow dance imitating, at a vastly slower pace, a three dimensional version of a boiling pot of water.   The causes of the motions are different, but the motions are similar.   It is from the turbulent mass of gas and dust that stars must form—but how?   Certainly, we know that the force which drives the atoms of gas and the grains of dust together to start the collapse into a proto—star is the force of gravity.   Each atom of gas and grain of dust is attracted to every other atom and grain.   The force of gravity tends to draw them all together.   As a cloud collapses, however, other forces are trying to push it apart.   Obviously, the turbulent motions of the cloud are trying to tear it to pieces.   But yet another force is trying to keep it from collapsing—gas pressure.

The pressure which a gas will exert on a container will try to prevent further compression.   The air in a balloon pushes outward on the inside wall of the balloon, keeping it from compressing the air further. Gravity, whether in a cloud in interstellar space or in a star, tries to act as a container, like the balloon.   And just as in a balloon, the pressure of the gas opposes it.   Pressure in a gas arises from the collisions of the gas molecules against each other and the walls of the container.   The collisions occur because the molecules are in motion, this motion being just another way of describing what we call "heat," or it's measure, temperature.   That is, the higher the temperature of a gas, the more rapidly the molecules move and the higher the pressure, all other things being equal.   The reverse is true: the lower the temperature, the lower the pressure—if you put a balloon in the freezer overnight, you'll find it collapsed the next morning. Leave it out to warm up, and watch it "blow itself up."

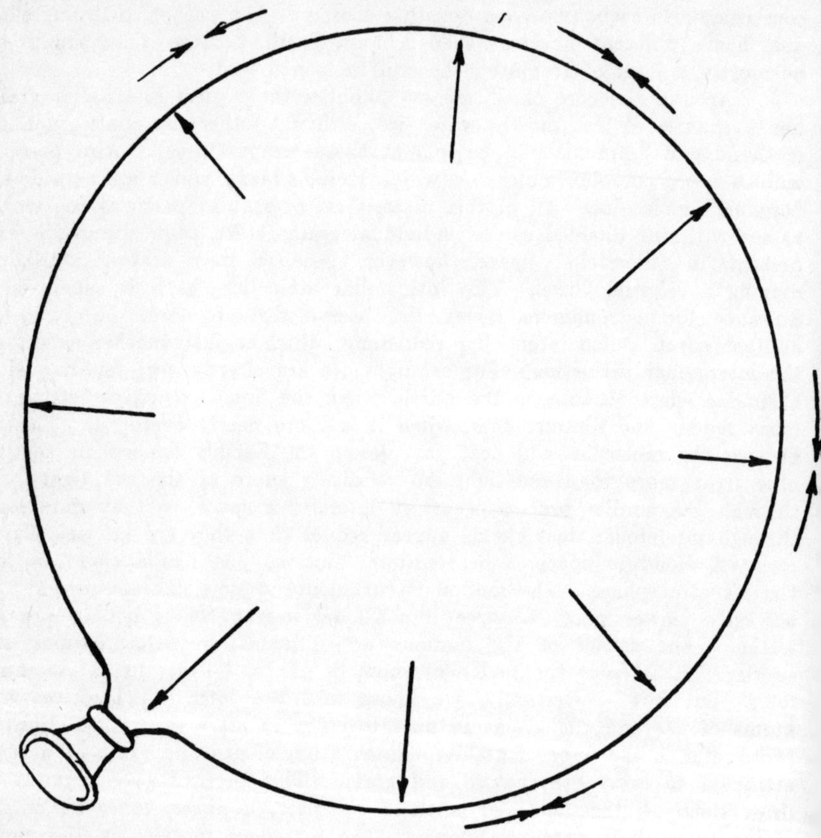

Figure 1.   The gas pressure of the gas in a balloon is opposed by the tension in the rubber.

A cloud of gas has a problem collapsing, because when it collapses, it heats up, which increases the gas pressure, which opposes the collapse. The fact that it heats up is something you can verify with the kind of air pump you use to pump up a bicycle tire by hand. If you try pumping up your bicycle tire, you will soon notice that the body of the pump heats up. Some of this heat is due to friction in the pump, but much of it is due to the fact that compressing a gas heats it up. So, the cloud has a problem, and it looks like it can't collapse. But wait! What if there were some way to cool the gas while the cloud is collapsing? Well, there is.   For example, the dust grains can radiate infrared light and radio radiation out of the cloud.   The molecules of the gas, which are moving because of the heat in the cloud, collide with the dust grains, giving them some of their energy.   The grains radiate the energy into space, and the gas is cooler than before.

Again, though, there is another "force" which tries to stop the collapse.   That is the inertial "force" which results from the rotational motion of the cloud.   In addition to the turbulent motion of the cloud, it is rotating.   This results from the chance collisions of the turbulent cloudlets in the cloud and its neighbors before the cloud begins to separate from the larger region of clouds.   As the cloud begins to collapse, it begins to rotate faster.   An example of this effect is that of the ice skater who does a spin.   Imagine the skater starting the spin with arms extended.   Then the skater pulls in his or her arms, and the rate of spin increases.   This is because the motion of the arms and hands, which is slow when they are further from the center of motion, become faster as they are brought in closer.   What has just been described is one aspect of what physicists call the conservation of angular momentum.   The fact that the rotation gets faster as the cloud collapses does not prevent the cloud from collapsing, but it does tend to make it flatten into a pancake instead of collapsing into a star.

With all the forces trying to prevent it, how does an interstellar cloud manage to collapse into a star at all?   Well, we are not actually sure.   It does happen, because we do see stars which we know to be very young.   And they are always found in massive clouds of gas and dust, and they are often found in groups near each other.   A beautiful nearby example is in the constellation of Orion, seen in the northern hemisphere's winter sky.   The young stars in the belt and sword are surrounded by nebulosity which is part of the cloud from which the stars have formed. In the sword is the Great Nebula in Orion, shown in Figure 2.   One thing we are pretty sure about is that a cloud probably never collapses to form a single star.   Young stars are generally found in groups, called clusters, which we think are formed from a large collapsing cloud.   Often smaller groups are found, right down to pairs of stars—binary stars. Many stars are found to be members of binary or multiple star systems— the Sun is apparently an example of the minority which are single. Multiple or binary star systems solve the problem of the conservation of angular momentum by putting much of the angular momentum into the motion of the stars about one another.   In the case of the Sun, the problem has been solved by putting much of the angular momentum into the motions of the planets around the Sun.   In any case, at this stage our theory is not sufficient to give the details of how the cloud starts its collapse, and how it breaks up into a cluster of stars.   One intriguing theory about how the collapse is triggered has the blast wave from a supernova explosion slamming into the cloud and compressing it until gravity wins over the forces trying to oppose it!

Once the cloud of gas starts to collapse, and gravity has won against its opponents, the cloud is on its way to becoming a star.   Again, typically the cloud is on its way to becoming a cluster, or a multiple star system.   Even in this case, however, at some point we can pick up the collapse after the cloud has broken into proto-stars, and follow just one of them.

Figure 2.   The Great Nebula in Orion.   This nebula is visible to the naked eye or in binoculars as the slightly fuzzy central star in the sword of Orion.

When a star forms out of an interstellar cloud of dust and gas, the dust grains which are caught up in the part which becomes the star are vaporized as the star is formed, so that the atoms in the dust contribute to the composition of the star.   Some dust is left over, and remains near the star.   It is thought that by some process, such dust, plus much gas, is the stuff from which planets like the Earth are formed.

As the cloud collapses, the central regions are compressed by gravity faster than the outer regions.   Thus, the density of the cloud increases faster in the center than elsewhere.   With this compression comes an increase in the temperature, and thus the pressure. As the gas heats up, infrared radiation tends to cool it so that gravity will continue to win. But only half of the energy given to the gas by the compression is radiated, so the rest heats up the gas. The fact that just half of the energy may be radiated follows from a physical law called the Virial Theorem, which describes the behavior of groups of particles which interact gravitationally. It would take a mathematical derivation to demonstrate the "half and half" split of the energy, but we can use this split to understand how the cloud evolves into a star. The result is a cloud which is collapsing faster the closer you are to the center, and which is heating up, again, faster the closer you are to the center. This process continues as the central regions draw together and get hotter and hotter. Much of the cloud is left behind, hiding the new star from our view until it is almost a main sequence star. It becomes a main sequence star after nuclear reactions converting hydrogen into helium are "ignited" in the center of the star and the star assumes a stable state. These nuclear reactions only occur at very high temperatures. And even at the high temperatures at which they occur, the rapidity with which they occur—the rate at which they release energy—becomes greater the higher the temperature. So, the collapse continues until the center of the star achieves temperatures above a few million degrees centigrade.

THE EVOLUTION OF SINGLE STARS

The ignition of nuclear reactions in the central regions of these stars halts the collapse and allows the star to become stable. *Hydrogen burning*, as this nuclear reaction is called, is quite efficient, and the star is made mostly of fuel—hydrogen. (Note that astronomers use the term "burning" with nuclear reactions, even though the process is not "burning" in the usual, chemical sense.) Further, stars which are on the main sequence are fainter than they will be at any time in their entire lifetimes until after they exhaust all possible nuclear fuel. Being fainter means that they use up their fuel at a slower rate. Thus, a star will stay on the main sequence longer than it will stay in any other evolutionary stage. As noted in the last chapter, the smaller the star, the longer will be its main sequence lifetime. Stars with a mass the same as the Sun's have main sequence lifetimes of almost 10 thousand—million years. Very small stars can have main sequence lifetimes far longer than the current age of the universe. On the other hand, a star of 100 stellar masses may have a main—sequence lifetime of only about a million years! Because the lifetime of a star on the main sequence is longer than in any other evolutionary stage, it is not surprising that if we look at a large number of stars, we will find most to be main sequence stars. That is the situation shown in the two HR diagrams in the last chapter.

The rate of the nuclear reactions is greater the higher the temperature, and the temperature increases as you go inward towards the center of the star. Thus, the nuclear reactions are strongly concentrated near the center. As the reactions proceed, a stage is reached when the hydrogen at the center has been converted entirely to helium. When this occurs, the reactions continue in the surrounding regions, and one has a helium core with a hydrogen *shell source* burning around it. The shell source continues to burn outward into the hydrogen—rich region, adding to the size of the helium core. No reactions occur in the helium core. In order to maintain the rate of the reactions, the star must arrange to heat up the regions where the shell source is burning, and it does so by shrinking slowly. In some stars, the gases in the core undergo turbulent bulk motions similar to the motion of the water in a boiling pot of water (but without the bubbles). This process is called convection. The convection will mix the material within a certain region, so that the hydrogen exhaustion will occur in the entire region. The same result eventually occurs: a helium core surrounded by a shell source.

Notice that the star, when it "needs" to heat up the core, "allows" the core to shrink. Another way to say this is that when the rate of reactions drops, the star is left with an energy deficit. That is, it is radiating more energy from the surface than it is generating in the interior. When that occurs, the gas in the star is cooled slightly, and the pressure drops. When the pressure drops, gravity wins slightly over its opponent, pressure, and shrinks the core. When this occurs, two things happen. Some energy is released from gravity itself, and half of it becomes available to make up part of the energy deficit. The other half goes into heating up the gas, which increases the nuclear reactions, which also makes more energy available to make up the energy deficit. You can see, then, that generating energy by nuclear reactions is a "temporary" (though long term) solution to a star's problem of gravitational collapse. Once the initial gas cloud starts its collapse, so that a star will be born, the long—term, inevitable process is one of gravitational collapse. Through its entire lifetime, the star has temporary methods of halting the collapse until the end—a place in the stellar graveyard.

When the slow shrinkage occurs in the core of the star, as hydrogen becomes exhausted in the center, and when the core becomes dominated by the heavier helium atoms, something paradoxical occurs. That "paradoxical" process is the expansion of the outer layers. The star, as seen from the outside, gets bigger. When this occurs, two characteristics of the appearance of the star can change: it can get brighter, and it can get redder. As it leaves the main sequence, it get brighter and redder. Then, for a while it stays at about the same brightness, while it gets redder. Then, depending upon the mass of the star, it may get brighter again. It now becomes a red giant (or supergiant, if it is very massive, see Figure 3). Since the star gets brighter, it burns what fuel it has left even faster than before. The end would come very quickly, except that the star has another possibility: the fusion of helium into carbon and oxygen. Whether or not this happens depends upon the mass of the star; the more massive it is the more assured we are that it will happen. This is because helium burning requires a much higher temperature than the ignition of hydrogen burning, and the more massive stars achieve higher temperatures in their cores. When helium burning begins, the shrinkage of the core is halted, and the star settles into another relatively stable state. Eventually it succeeds in consuming all of its helium, and the shrinkage of the core resumes.

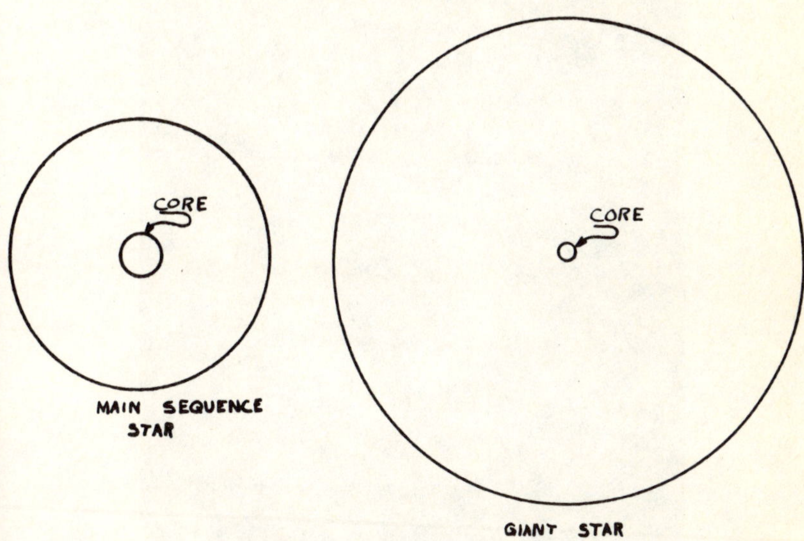

Figure 3.   The relative sizes of the core and surface of the star are shown for main sequence (left) and giant (right) stars.

The continued process of burning the "waste" of the previous stage of nuclear burning so that it becomes the "fuel" for the succeeding stage can continue through several stages if the star is sufficiently massive. Each succeeding stage requires a higher temperature, so the more massive the star, the further it can go. The Sun, when it has evolved sufficiently, will be able to burn helium into carbon and oxygen, but that will be the end of the process for it. Stars of more than about eleven solar masses will be able to burn successive species of nuclei until ^{56}Fe (iron) is produced. The iron atom consists of a nucleus with 26 protons and 30 neutrons, surrounded by a cloud of 26 electrons. In the center of a star, the electrons have all been stripped off, and are present as an all—pervading "electron gas." The ^{56}Fe nucleus is the end of the line for nuclear fusion reactions. That is, attempting to build heavier nuclei, such as cobalt, nickel, copper or zinc by adding protons or helium nuclei to the iron nucleus would require putting energy in, rather than resulting in energy release. Thus, once a massive star has produced an iron core, it can no longer produce energy by nuclear fusion in its core.

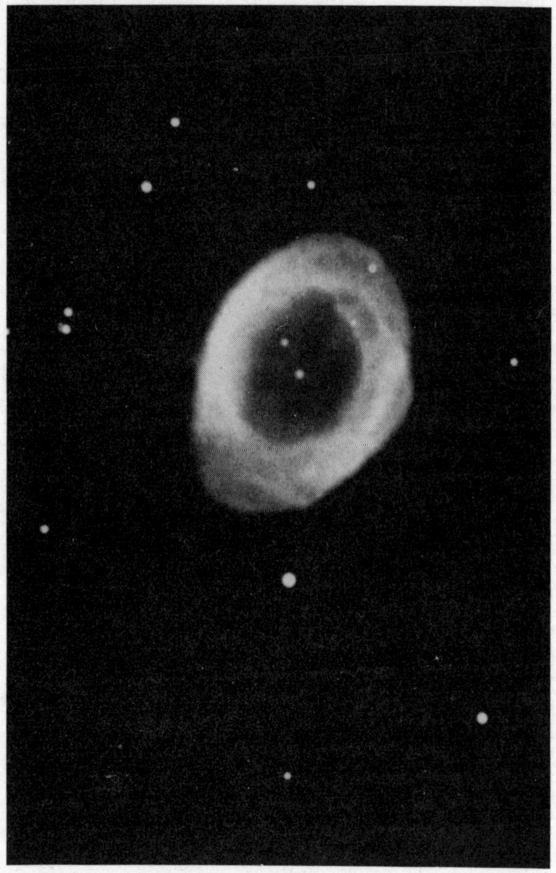

Figure 4.   The Ring Nebula, a planetary nebula in the constellation of Lyra.

Throughout the series of events described above, the outer layers of the star retain a composition roughly the same as the original gas cloud from which the star formed. This is because there is, at most, only minor amounts of mixing of core material outward. Thus, for much of the life of the star, the composition we measure by analyzing the star's light remains the same, even though there are drastic changes in the star's core. The exceptions are minor but important. During some of the successive nuclear burning stages a small amount of material from the outer edge of the core can be mixed outward, and be seen as a contaminant at the surface. Such evidence can be important for testing the accuracy of our models of stellar evolution, since the model must be substantially correct if it can accurately predict the amount of the contaminants.

Finally, during the late evolutionary stages of stars, it is common for some of the outer layers of the star to be lost into space. This phenomenon, which is called *mass loss*, can happen in any of several ways. The actual physical processes are not well understood in any of them. The consequence, however, is that a star in the late stages of its evolution can be found with only a small part of its hydrogen envelope remaining.

The case of a star with a mass about the same as the Sun's is important: after the star has exhausted helium, it cannot ignite carbon burning, and it must undergo further shrinkage of the core. As the core shrinks, the outer layers are expelled into space, and form what is known as a planetary nebula. A planetary nebula is so named because it appears like a disc resembling a planet in a small telescope. The Ring Nebula in the constellation of Lyra (see Figure 4) is a beautiful example. A planetary nebula is actually a shell or "bubble" of gas surrounding the star, and expanding outward. The star in the center of the nebula is essentially all carbon and oxygen, and it is no longer capable of nuclear reactions. The remainder of its evolutionary life–cycle is a story of slowly cooling off, ending its life as a *white dwarf*. White dwarfs will be discussed in the next chapter.

THE EVOLUTION OF BINARY STARS

The discussion given in the preceding sections is somewhat misleading, because it refers to the case where the star is alone in space, without any interaction with another star. But most of the stars in the neighborhood of the Sun are members of binary or multiple star systems. In binary star systems, the two stars revolve around each other. The most commonly observed binary systems have two stars of nearly the same mass, but systems in which the two stars have very different masses are also not unusual. In most systems, the stars revolve in nearly circular orbits, which means that the distance between the two stars remains nearly the same as they move around their orbits. We find systems in which the two stars are so close they literally touch each other, and we find other systems in which the stars are separated by truly enormous distances. As you might expect, in those binaries in which the separation is large, we expect the evolution to proceed as if each were a single star. But in those binaries in which the two stars are touching, their evolution proceeds in ways which can be very different from that of single stars.

The reason that the evolution is different in close binaries is that at certain stages one star may lose material from its outer envelope and be transferred to the other star. This changes the situation, because the loser is now a smaller mass star than before; it is also important to realize that it loses this mass from the envelope only. Consequently, the other star now becomes more massive than before, which changes its evolutionary cycle. Finally, it can easily happen that one of the stars loses material which escapes the system entirely, travelling out into space. This changes the evolutionary cycle of the loser, and it will also change the orbit of the system.

Imagine that you are in a spaceship travelling in the vicinity of a binary star system. We'll ignore the effects of gravity on your spaceship, and assume that the heat from the stars does not have any effects on the ship or the people inside. Think about what would happen to a small particle, a baseball for example, if you were to let it fall out of the spaceship some distance from the "surface" of one of the stars. You might think that it would fall to the surface of the star, but it is more likely that it would orbit the star, a short distance above the "surface." (This is because your ship would be moving in an orbit, and the baseball would have the same motion as the ship.) If you are "close enough" to the star, the effects of the other star will be so small that they can be ignored. The baseball can be said to "belong" to the first star. If we were to move the ship to an orbit around the second star, we would have the same result if we were "close enough;" the baseball would "belong" to the second star, and would orbit about it. But, what if you were to let the baseball go at some position in between the two stars. What then? Suppose the two stars had the same mass, and you placed the baseball exactly halfway between the two stars. In principle, the baseball would sit there, falling towards neither one nor the other star. In practice, it would not take much of a disturbance to start the baseball falling in the direction of one or the other. This point is called the *First Lagrangian Point*.

Suppose you draw a line from the center of one star to the center of the other, and suppose you test which star the baseball belongs to at various positions along the line. It would be found to belong unambiguously to one or the other. What about positions which are not on the line between the stars? At some of these, neither star will start to dominate. It turns out that, as one goes up from the surface of one of the stars in any arbitrary direction, one eventually gets to a point where the baseball no longer can be said to "belong" to that star. That is, outside that point, that star does not dominate gravitationally; if let go, the baseball will either be dominated by the other star, or it will be lost from both stars. The edge of the region in which a given star dominates is a mathematical surface called the *Roche lobe*, after the French astronomer who worked out the theory. Thus, each star has a Roche lobe, and they touch at the first Lagrangian point.

The Roche lobe is important in the evolution of a star in a binary system. Suppose one of the stars is somewhat more massive than the other, and it, therefore, evolves through the main—sequence stage more quickly than the other. It becomes a red giant while the other is still a main sequence star. As it evolves into the red giant stage, the outer layers swell outward. For a single star, there is no limit placed upon this process, except that of the star's own internal structure. But, what if a star swells up so that it fills its Roche lobe? The result is that the outer layers of the star will no longer "belong" to the star gravitationally, and will drift off, either into space around the binary system, or onto the other star. How much is lost in space and how much goes to the other star depends upon the circumstances. The "circumstances" include the vital question of how big the Roche lobe is compared to how large the star would become if the other star were not present. They also include the rapidity with which the star is expanding. A star which is expanding rapidly to a size which would be much larger than the Roche lobe will end up losing a lot of its mass. Some of the mass will be lost into space, and much of it will end up in a "disk" of material in the orbital plane; ultimately much of it will end up on the surface of the formerly smaller star. "Formerly" is the key word here; this process can result in the hierarchy of the two stars being reversed. That is, the one which started out as the most massive can become the least massive, and vice versa.

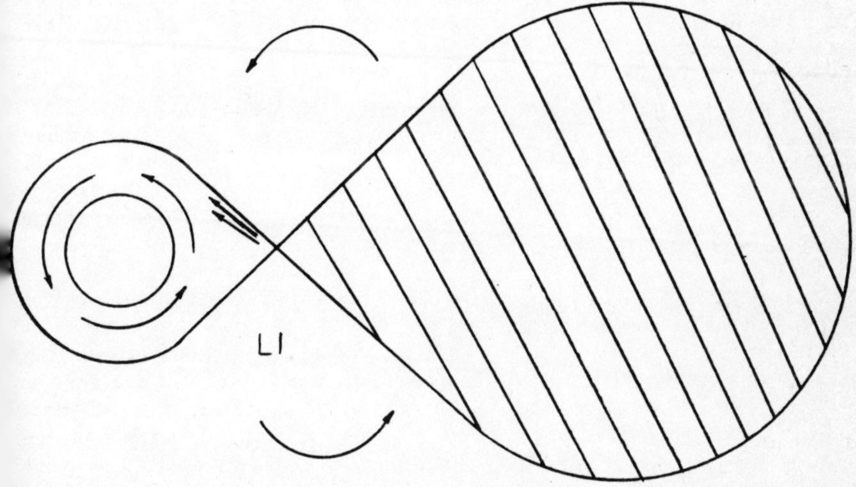

Figure 5.   A binary star system in which one star fills its Roche lobe. The first Lagrangian point is also marked.

If a star loses its outer layers by this process, its evolution will be completely different than if it had retained it. It may not be able to ignite the next stage of nuclear burning, even though it might well have if had been a single star.

CHAPTER 9

THE STELLAR GRAVEYARD

WHITE DWARFS

When, at some point in its lifetime, a star cannot begin the next phase in the sequence of nuclear burning, what happens?   Well, what happens then depends upon how massive the star is.   If a star has a mass in the neighborhood of one solar mass, it will have a core of carbon and oxygen.   After the red giant phase, in which it will have "burned" helium into carbon and oxygen, it will "blow off" the outer parts of its envelope, leaving a planetary nebula and a small, hot star which is almost entirely made up of the former core of the star.   Typically, the remaining mass will be about 0.6 times the mass of the Sun.   This object is no longer generating nuclear energy, but it is still radiating energy from its surface.   As a result, it must cool off and shrink.   Eventually, the interior regions attain a state called *degeneracy*, one of the more exotic states of matter known, and the star becomes a white dwarf.

In order to explain degeneracy, we must back up a bit, and talk about atoms, their nuclei, and electrons.   We said before that the nucleus of an atom includes protons and neutrons, with the number of neutrons equaling or somewhat exceeding the number of protons.   The number of protons is what determines the type of chemical element.   Normally, the nucleus is surrounded by a "cloud" of electrons, whose number equals the number of protons.   The electrons are easily detached, especially the outer ones.   In the case of an atom like oxygen, with eight electrons, the outer two are pretty easy to pull off.   If the gas gets very hot, the outer electrons will get stripped off when the atom collides with another.

In any hot gas, then, one finds atoms which are missing a few electrons; these atoms are called ions.   In the space between the ions is a sea of free electrons.   The sea is being joined continually by electrons newly stripped off of some atom, but it is continually losing electrons which are captured in collisions with ions. As the temperature and density increase, deeper and deeper layers of electrons are stripped off the atoms. In a red giant which is burning helium into carbon and oxygen the temperature and density in the core are such that all of the electrons have been stripped off, leaving bare nuclei swimming in a sea of electrons.

Once the nuclear burning stops, the density in the core begins to increase, as the star shrinks. Eventually, a point is reached where something strange happens; the electrons get pushed into a state where they can't be compressed any more. That is, the electron gas begins to provide a greater pressure than the ion gas. The compressibility of the electron gas is determined by the quantum—mechanical properties of the electrons, and not by the temperature, as is true for normal gases. When this happens, the size of the star is no longer determined by how hot the interior is. That is, in a star which is not degenerate, with no nuclear reactions going on, the central regions of the star will shrink as it radiates energy from its outer layers. This process could go on practically indefinitely if degeneracy didn't occur. In stars which do become degenerate, the shrinkage is halted by the electron gas, and the star settles into a configuration in which it stays the same size as it cools off. It is now a white dwarf, and its density is incredibly high. A white dwarf with a mass roughly the same as the Sun's would have a size roughly the same as the Earth's, and its density would be hundreds of thousands of times greater than the densest materials we find at the surface of the Earth.

The physical structure of white dwarfs was worked out by the Indian—born astrophysicist S. Chandrasekhar; in recognition of his contributions to many fields of physics (most of which are important in astronomy) he was awarded the Nobel Prize (jointly with W. Fowler) in 1983.

By 1935, Chandrasekhar had worked out the structure of white dwarfs, but it was several years before his proposals were fully accepted by the community of astronomers and physicists. The theory which he published has some startling qualities, and one of the most remarkable concerns the sizes of white dwarfs of different masses. As described above, once a star has achieved degeneracy in the later stages of its evolution, it settles into a state where it cools off at a constant radius, or size. The theory predicts that stars of the same mass and the same composition will have the same size. What is remarkable is that it predicts that the more massive the star, the *smaller* it will be. This is just the opposite of the case of main sequence stars. The more massive the degenerate star, the more the gravitational force on the layers of the star squeezes the electron gas, and the most stable state is a smaller star. The theory further predicts that at a certain mass the star will have a zero radius! What this means is that stars above the certain mass simply cannot exist as white dwarfs; they must shrink into neutron stars or black holes, which we will discuss in the next sections.

Figure 1.   Nobel prize winner S. Chandrasekhar.

In any case, the theory says that we will not find any white dwarfs above a certain mass, and that mass will depend upon the composition of the white dwarf. For the compositions we are concerned with, that mass will be about 1.4 solar masses. Now, the theory of stellar evolution predicts that stars up to around eight solar masses could evolve into white dwarfs, except that no white dwarfs can be that massive. Do they make something else?   No, it appears that they push off enough mass while in the red giant stage to get the remainder under 1.4 solar masses. Thus, most stars up to a few solar masses insist upon becoming white dwarfs, and arrange their affairs while red giants so that this will occur.

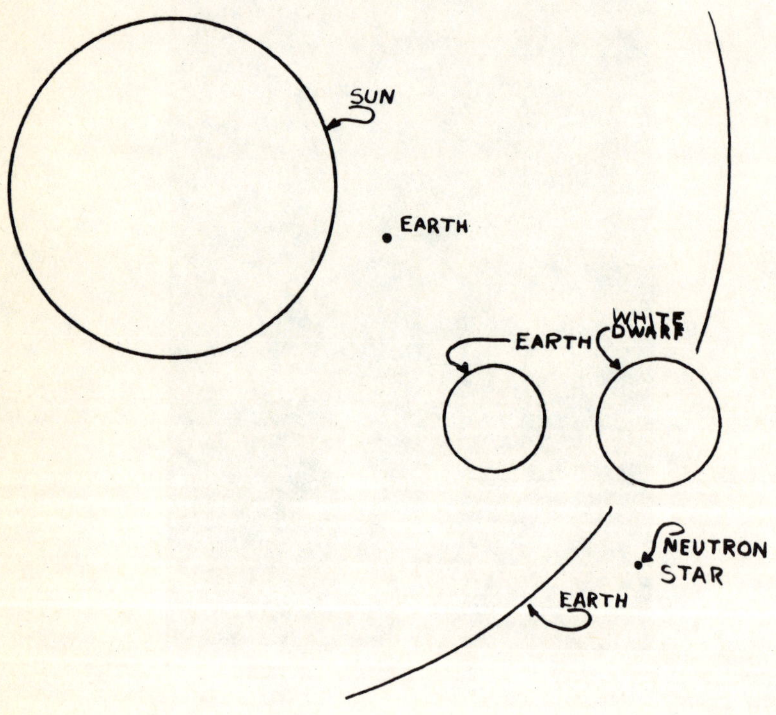

Figure 2.   The relative sizes of the Sun, the Earth, a white dwarf, and a neutron star are shown.

Let us take a bit of a diversion, at this point, and talk about white dwarfs in binary systems. In the preceding section, we considered a binary system in which mass exchange had reversed the hierarchy of the two stars, leaving the formerly more massive star with a helium core and a much reduced envelope. The formerly less massive star is now evolving more rapidly, and will soon leave the main sequence. The helium—core star now has no way of generating energy except for a shell source burning hydrogen into helium in the layers surrounding the helium core. As the star evolves, the hydrogen—rich envelope becomes thinner and thinner, while the helium core grows slowly at its expense. Eventually, the amount of energy which can be obtained in this way becomes less than the star radiates, and it begins to cool off. By this time, the central regions of the core have become degenerate, so the star is becoming a white dwarf.

The scenario described above is not the only way to produce a close binary containing a white dwarf, but it is one method which is thought to be common. We do see evidence for close binary systems containing white dwarfs, as will be described below. Thus, this process and some others must be occurring with real stars. What happens next? Well, remember that we have the old main—sequence star evolving rapidly, and it soon exhausts the hydrogen in its core. It begins to swell up, as it starts evolving into the red giant stage. But just as was true for its formerly more massive companion, it soon finds itself filling its Roche lobe, and spilling envelope material into space, and onto its white dwarf companion.

What happens when hydrogen—rich material falls onto the surface of a hot helium white dwarf? Remember that hydrogen is the "fuel" of the nuclear reactions in most stars. For awhile, nothing happens. The hydrogen builds up on the surface, making a layer of hydrogen on top of the helium. One does not have to go very far down into the white dwarf to find temperatures high enough to start hydrogen burning, and very high densities as a result of the very high gravity at the surface of a white dwarf. After awhile, on the order of 10,000 to 100,000 years, enough hydrogen has built up to produce explosive conditions. An explosion does result, an explosion which is essentially a big hydrogen bomb going off. For a short period of time, the brightness of the star, as seen from a distance, has increased by 10,000 times. This explosion is called a nova. Violent as it is, it is not violent enough to disrupt either star or the binary system. In fact, after the explosion has occurred, the larger star will just resume dumping material on the white dwarf, and the stage will be set for a "repeat performance." When the conditions are such that we observe repeated explosions, the nova is called a recurrent nova.

NEUTRON STARS

The next object in the stellar "zoo" is the *neutron star*. The theory of the neutron star was worked out by the American physicist J. Robert Oppenheimer (1904—1967) and his colleagues in the 1930's as a purely theoretical exercise; there was no evidence at all that such things existed or could exist. (A few white dwarfs were actually known before their theory was worked out.) In a white dwarf, the sea of electrons has been squeezed down into the smallest configuration it can achieve. How could you squeeze the gas into a smaller space? The only answer appears to be to get rid of the electrons. If you squeeze the electrons into the protons to form neutrons, the resulting "neutron gas" will occupy an even smaller volume. A star made up of a degenerate neutron gas will be at densities similar to, or exceeding, that of the nuclei of atoms; a neutron star the mass of the Sun would be only a few kilometers across. Neutron stars, like white dwarfs, have a maximum mass, but theirs is larger than the maximum for white dwarfs. The maximum mass of neutron stars is about two to three solar masses, though this value is very uncertain.

How are neutron stars formed? Because it takes a lot of energy to squeeze electrons into protons, to produce neutrons it appears that they can be formed only in supernova explosions. Current thought excludes any continuous non—violent processes such as characterize the formation of white dwarfs. This conclusion appears to be reliable even though only a few of the known neutron stars are associated with *supernova remnants* (see Chapter 11). Neutron stars are seen as *pulsars*; which also will be discussed in Chapter 11. Hundreds of pulsars are known; most are not associated with supernova remnants, but their statistics are consistent with their having been formed in supernovae. Could you form a neutron star by piling mass on to a white dwarf, as suggested earlier? Evidently the answer is no; it simply blows up, leaving nothing behind.

BLACK HOLES

There is one more occupant of the stellar zoo, which isn't "stellar" in the usual sense. It is the end result when stars collapse catastrophically—the *black hole*. Black holes are known by what they are not rather than what they are. What they are not is sources of light; nothing, including light, can escape from a black hole. On the other hand, light and anything else can fall in—hence the name black hole. The first steps to a theory of black holes were made by the German astronomer Karl Schwarzschild in 1915—16. The theory was more fully worked out by J. Robert Oppenheimer and his colleagues in the 1930's. Again, it was a purely theoretical exercise. What the theory says is that it is possible for matter to be compressed to such high densities that the gravity becomes so strong that light cannot escape from the object. In the region defined to be the black hole, the structure of space—time is distorted to such a degree, compared to the general outside universe, that our ordinary laws of physics don't apply and Einstein's general theory of relativity must be used to describe the situation.

It may seem incredible that light cannot escape from the black hole; after all, isn't light faster than anything else? In order to get a grasp on the situation, let's try the following "thought—experiment." How fast do you have to throw a baseball in order to have it escape completely from the gravity of the Earth? If you've tried throwing a baseball upward, you know you can't even get close to making it escape. So, you try a cannon, to shoot the ball up. Even this probably wouldn't do it. Given sufficient determination and technology, though, you would probably find a way to get the ball going fast enough, about 11 km/sec, that it would escape. What about on the Moon? Remember, for a moment, the Apollo flights to the Moon; you know that the gravity at the surface of the Moon is about one-sixth that at the surface of the Earth so the astronauts "hopped" or "skipped" rather than walked. It turns out that a ball would have to be launched upward at a speed of only 2.4 km/sec to escape the Moon. What determines the escape velocity of an object like the Earth or the Moon? Only two characteristics are important, the mass of the object, and its radius (assuming we are talking about the velocity of escape from the surface). The Moon has a smaller mass which results in a smaller velocity of escape. The moon also has a smaller radius, which would result in a larger velocity of escape. In the case of the moon, the smaller mass is more important, and the net effect is a smaller velocity of escape. What about the Sun? The mass of the Sun is much larger than the Earth's, but so is its radius. In this case, the escape velocity is much larger, about 620 km/sec! If we were to shrink the Sun, keeping the mass constant, down to the size of the Earth, the escape velocity would be ten times as large as the Sun's is now. This approximates the case of a one solar mass white dwarf. Suppose we shrink the Sun down to a diameter of 20 kilometers, the size of a neutron star: The escape velocity from the surface will be about half the speed of light!

One can see from the preceding paragraph that one solar mass of material needs to be compressed down to a size not much smaller than a neutron star for the escape velocity to equal the speed of light. So far, we have been talking about theory; the question of the reality of black holes and how to observe them must be dealt with. If nothing can escape from a black hole, how can we detect one? There are two ways in which a black hole makes itself known, and together they can be used to make a case for its existence. The first we have mentioned already: it has a gravitational field. If the black hole is a member of a binary system, the motion of the companion will be the same as if the black hole were a star having the same mass. This, we can determine its presence and its mass with suitable observations of the companion star. The other way the black hole makes itself known is the result of what happens when material falls into it. Because of its small size, whatever falls in must be compressed to very high densities. Material compressed to very high densities very quickly heats up and radiates. In this case, the material will be heated to very high temperatures, so we will see it radiating not only in visible light, but also ultraviolet and even X-ray radiation. The radiation we see occurs just before the material disappears into the black hole, so we are not really seeing the black hole. The intense source of ultraviolet and X-ray radiation is characteristic of material being compressed to very high densities. There is only one other object which can cause this same phenomenon: a neutron star. Material falling onto the surface of a neutron star is also compressed to very high densities, and the radiation pattern is similar to the case of the black hole.

In order, then, to detect a black hole, we must find a source of ultraviolet and X—ray radiation, but we must also distinguish between black holes and neutron stars. If the black hole is in a binary system, though, we may be able to determine its mass from the motion of its companion. If the mass we get is too large to be a neutron star, we must have a black hole. This is just the technique which has been used to select the three candidates we now have for stellar black holes: Cyg X—1, LMC X—3 and A 0620—00. All three are X—ray sources located in binary systems; the first two contain a normal B—type star and the latter contains a normal K—type dwarf. The observations indicate masses of about 10 to 15 solar masses for the "unseen companion" in Cyg X—1, about 6 to 14 solar masses in LMC X—3, and at least 3.2 solar masses in A 0620—00. If these values are to be trusted, they are above the limits we think hold for neutron stars. Thus, they are probably black holes. We can't be sure, though, because the observations are difficult to make, and there may be other explanations for what we see.

If black holes really do exist, we must explain how they can be created. The only mechanism we can imagine which could produce stellar—sized black holes (as opposed to the hypothetical black holes in the centers of galaxies or in quasars, which are another matter entirely) is a supernova explosion. It is to this "ultimate of stellar exhibitionism" that we now turn.

CHAPTER 10

THE SUPERNOVA EXPLOSION

TYPES OF SUPERNOVAE

The basic features of supernovae were described in Chapter 3. Let's review some of them here. What we see is a star after it has brightened in a very short time to a luminosity about a thousand—million times that of our Sun. The supernova stays near its maximum brightness for a matter of months, and then slowly dims. Two or more years may pass before it disappears. For a month or so, the luminosity may rival the entire galaxy in which it is located. Afterwards, it leaves behind a ball of gas which is found to be expanding at very high speeds, as in the case of the Crab Nebula. In general, supernova explosions have considerable similarity to one another. But, if one looks closely at descriptions of all the supernovae which have been observed, one finds that there are differences, and that they can be classified into groups. Supernovae can be grouped into Type I and Type II. At various times, people have suggested additional types; the prolific investigator and pioneer of the field, Fritz Zwicky, listed five types. At present, most investigators seem to accept the two main types, with some accepting a third. The two types differ in the following important ways:

1. Light Curves: The Type I supernovae are the brighter of the two, becoming two to three times as bright, on the average, as Type IIs. They stay bright for a shorter time, however, dropping off quickly after a month or so. The Type II supernovae drop off less rapidly at first, but after approximately 100 days start dropping more rapidly. The Type I, after the initial rapid drop, fall less rapidly. An important difference between the two groups is their homogeneity: a very high percentage of the Type I supernovae behave much like each other, whereas the Type II supernovae differ, from one another, to a much greater extent.

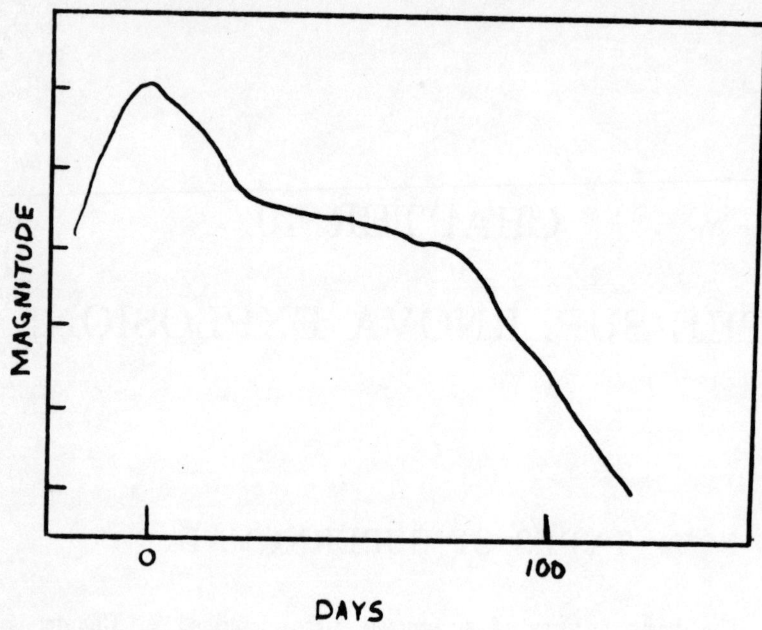

Figure 1.   The mean light curve for Type I supernovae.

2. Spectra: When the spectra are examined, the Type II supernovae show lines of hydrogen, whereas the Type I do not.   Since hydrogen is the most common element in normal stars, this difference is quite significant. The presence or absence of hydrogen in the spectrum is, perhaps, the most secure way to distinguish between Type I and Type II supernovae, because the light curves show enough variations to make classification difficult at times.

3. Location in galaxies, and types of galaxy in which they are found: Our Milky Way galaxy is one of a type called a spiral galaxy.   The stars in galaxies can be grouped into two types: Population I and Population II. Population I stars, of which our Sun is an example, are a younger population of stars, including those which are being formed currently. This population is comprised of stars of all masses, including very massive stars.   Population I stars generally are found in the spiral arms of spiral galaxies, and in all parts of irregular galaxies.   Population II stars are an older population of stars, thought to be formed shortly after the galaxy— well before Population I stars.   In spirals, they occupy the central bulge and the halo, and include stars in the globular clusters.   In our galaxy, no new Population II stars have been formed for 10 thousand—million years, and all the high—mass Population II stars have already evolved to

their death.   Thus, Population II stars now just include stars of rather small mass.   In elliptical galaxies, essentially all of the stars are apparently analogous to the Population II stars in our galaxy.   The location of the two types of supernova seem to divide according to population in the following way: Type II supernovae seem to be closely associated with Population I stars (yes, that is not a misprint—nobody said that the labels of these classifications had to make sense!) whereas Type I supernovae seem to be found everywhere.   That is, it appears that Type I supernovae can be of either population type.

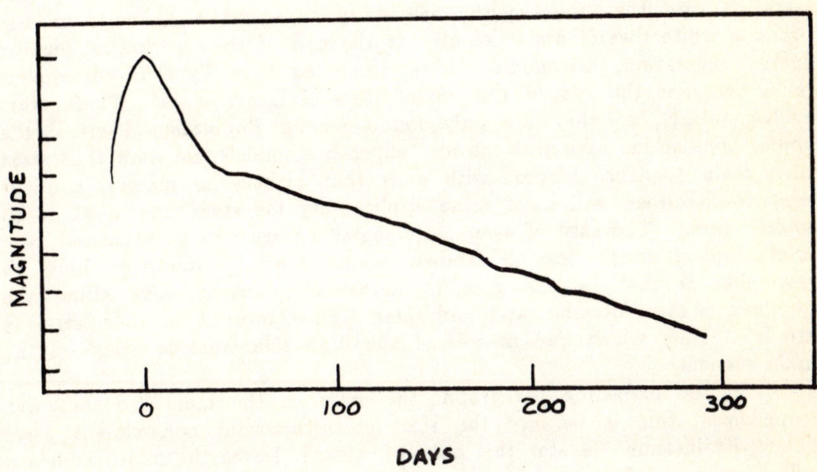

Figure 2.   The mean light curve for Type II supernovae.

4. Frequency: This is a characteristic in which the two types seem to be similar in a galaxy like our own; the statistics indicate that both types occur at a rate of about one every 40 years.   Of course, there are no Type II supernovae in ellipticals.

MODELS OF THE SUPERNOVA EXPLOSION

After a few decades of attempts to model the supernova explosions, it has become clear that the two types of supernovae involve very different types of pre—supernova objects.   The only similarity seems to be in the amount of energy released in the explosion, and the fact that nuclear reactions occur which produce large quantities of heavy elements. Supernova 1987A has been classified as a Type II supernova, so we will spend most of our time on this type.

TYPE II SUPERNOVAE

The characteristics of Type II supernovae, which were described above, indicate that they are associated with a population of stars which includes very young stars. Since stellar evolution calculations for normal stars indicate that those with masses up to about eight solar masses produce white dwarfs non—violently at the end of their period of nuclear energy generation, astronomers have concluded that Type II supernovae must occur at the end of the life of stars of larger mass. These stars evolve quickly, so they are only found among Population I stars: the young population. We will consider supernova models for stars that start their main—sequence lifetime with more than eight solar masses, and the scenario described will have some applicability for stars up to 30 or 40 solar masses. The case of even more massive stars will be discussed very briefly later; much less is known about them. Another thing to remember is that as one goes to more and more massive stars, the numbers of stars become rarer and rarer. Thus, most Type II supernovae are from stars which had masses of about 15 solar masses when on the main sequence.

In the preceding paragraph, the mass on the main sequence was emphasized; this is because the star looses mass in non—violent ways during its lifetime. A star this massive, after it leaves the main sequence, will become a red supergiant. It will first be burning helium in its core. During this and successive red supergiant phases, it will be continuously losing mass from its outer atmosphere. The envelope of the star has swelled out greatly in response to the shrinkage of the core after hydrogen exhaustion and it does not take much energy to drive mass away from this greatly distended outer envelope. The physical mechanism of this process is not fully understood, however, so we cannot give a detailed description of the process here. We do observe "circumstellar shells" around red supergiants, however, so mass loss is well documented. The consequence of the mass loss is that the star can lose a few solar masses of hydrogen—rich material before it gets to the point where a supernova explosion is imminent. The presence of this cloud of gas surrounding the star can affect the light curve of the supernova, because the radiation and the ejected gases will slam into it early in the explosion.

As the star burns its helium, it produces carbon and oxygen in the core. During this time, a hydrogen shell source is producing more helium at the outer edge of the helium core. This is the beginning of an "onion" effect in which the core becomes a set of concentric shells of differing composition. When the helium is exhausted in the central regions, the star faces the problem of producing enough energy to make up for the prodigious amounts radiated from its surface. The answer, as before, is gravitational shrinkage of the core, while the envelope becomes even more distended. As the core shrinks, energy is released, half of which is radiated, and half of which goes into heating up the core. Eventually carbon is ignited, and another phase of nuclear energy release starts. Oxygen also will burn during this stage, and a core of silicon and other heavy elements is produced. Another layer is added to the onion. If the star is sufficiently massive, the silicon may ignite and burn to produce iron and nickel.

In the Chapter 8 we noted that nuclear energy generation could only proceed as far as iron. When the star has made a core of iron (and some nickel, which is produced along with the iron), it can not produce any more energy by making heavier elements. In order to have more energy to radiate, the star does what it has done many times before; it takes energy from gravity. The core shrinks, and the outer envelope expands a little more. Also, as before, the core heats up a bit. Additionally, during this phase there are shell sources burning at the outer edge of each layer of the onion. But these shell sources can't produce enough energy to solve the star's problem.

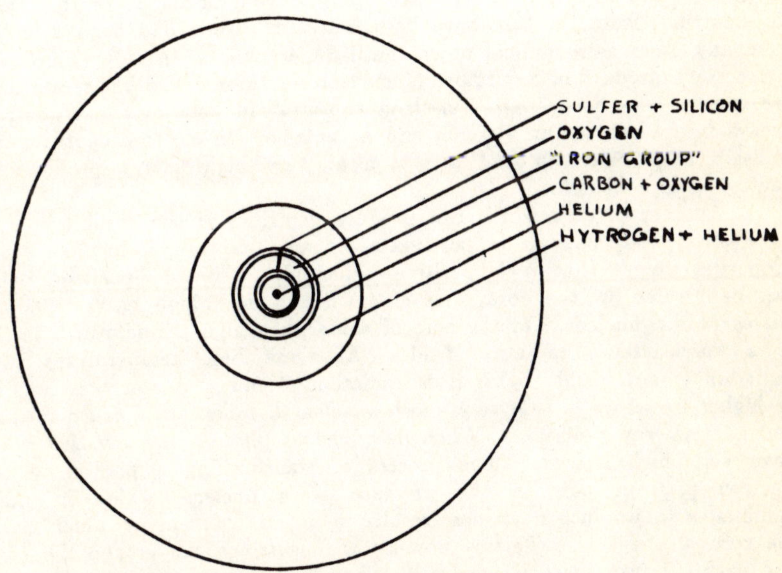

Figure 3. The "onion layer" model of a massive star late in its life is shown. The outer layer, labelled "hydrogen and helium," has a chemical composition little changed from when the star was first formed.

It is good to remember that what holds up the layers of the star against the pull of gravity is the pressure of the gas. The core is intensely hot, now—on the order of a thousand—million degrees. It is the pressure of this hot iron—nickel gas, which is caused by the collisions of iron and nickel nuclei as they race around at the high speeds reflecting the high temperature, which prevents collapse. Once the temperature gets high enough, however, some processes begin to occur which spell doom for the star. These are processes which take away energy from the core; energy which is needed to maintain the high temperature and the high pressure. One of these is the production of neutrinos.

A neutrino is an elementary particle. Unlike electrons, protons, and neutrons, it is not one of the "building blocks" of atoms. Unlike the photon, which is the way electromagnetic radiation manifests itself as a particle, the neutrino is very hard to detect. Neutrinos are produced in nuclear reactions, and the reactions in the center of the Sun are flooding our bodies with neutrinos continuously. During the time it takes to read this paragraph, thousands of millions of neutrinos from the Sun have passed through your body. Neutrinos are so elusive that they pass through the Earth without interaction. They are detected only because so many are produced that even with their extremely low rate of interaction, a few are still detected. They were first described theoretically by Wolfgang Pauli in 1930. They have been detected in the laboratory, and the neutrinos from the Sun have been detected. None had been detected from any other astronomical object until Supernova 1987A. Neutrinos are a necessary product of a number of nuclear reactions. The basic reactions are called "beta decay" and "electron capture." In beta decay, a neutron decays into a proton, an electron and a neutrino. In electron capture, the collision of an electron and a proton at very high energy produces a neutron plus a neutrino.

Another process which can produce neutrinos is the annihilation of an electron—positron pair (a positron is an anti—electron; it has the same properties except that it is positively charged). Electron—positron pairs can be created in the core of a star when a very high energy photon passes near a nucleus. In the core of a star, the high temperatures result in a very intense radiation field. At these high temperatures, the radiation is not visible light; it is radiation of much shorter wavelength, or higher frequency. This very short wavelength radiation is called X—ray and gamma—ray radiation. When described as photons, we say that they have very high energy. The process of transforming a photon to an electron—positron pair in the presence of a nucleus, which in time annihilates to produce neutrinos, results in a "leak" of energy which cools the core of the star. This is because the neutrinos can escape the core and pass out into space immediately, whereas the photons take millions of years to work their way out. Later, after the core starts collapsing, the neutrinos get "bottled up" for a few seconds, but by then it is too late. This process, pair annihilation, is actually the most important mechanism of energy loss from stars in stages of evolution following helium burning. It is more important than loss by electromagnetic radiation from the surface of the star.   The magnitude of the neutrino loss increases with each stage, and this loss speeds up the rate of evolution.

As the core shrinks and heats up, another process begins to occur, called "photodisintegration."   What this means is that the photons of high—energy electromagnetic radiation start breaking up the iron and nickel nuclei.   That's right—the nuclei which the star spent millions of years putting together now get torn apart.   It takes energy to tear them apart, so this process is another "leak" of energy which cools the core.   A lot of energy can be soaked up splitting the nuclei into helium nuclei and even protons and neutrons.   Once protons start to appear, another process adds to the problem—electron capture.   Electrons and protons combine to form neutrons.   This is a two—fold disaster: neutrinos are produced, which carry away energy, **and** electrons are consumed.   The electron gas has been an important contributor to the gas pressure of the core, and when the electrons begin to disappear, the pressure begins to drop.

Once these processes have started, the end is suddenly at hand. The iron and nickel nuclei are broken down and this plus electron capture results in the production of neutrons.   In a time which is a very small fraction of a second, the core of the star, with over a solar mass of iron and nickel, collapses.   Suddenly, there is nothing "holding up" the outer layers of the star.   The inner core collapses into a neutron star, and the outer core begins to rain down upon it.   What happens next is the subject of considerable controversy.   An enormous amount of gravitational and nuclear energy is released, more than enough to power a supernova explosion.   The problem is that all the mass is falling inward, not outward.   Furthermore,   much   of   the   energy   is   going   into photodisintegration of the outer core.

The problem of how to produce an explosion from a collapse was dealt with in an imaginative way in the models by Stirling Colgate and his collaborators in the early 1960's.   Their hypothesis was that the neutrinos carried the energy out from the inner core to the outer core. That is, the intense flow of neutrinos produced in the inner core would escape, because the neutrinos could hardly interact with the neutrons of the inner core.   When the neutrinos reached the outer core layers, however, enough would be absorbed, even with their very low chances of interaction with ordinary nuclei, to power the explosion.   Subsequent calculations show that this model had some difficulties as originally proposed, but its publication stimulated much thought about supernovae in succeeding years.

The most common scenario now advanced is that the collapse of the core ends with a "bounce"—the infalling gases from the outer core hit the surface of the neutron star and rebound.   This produces a shock wave, which pushes outward through the star, heating and expanding the outer   layers.   Much   of   the   energy   in   the   shock   goes   into nucleosynthesis—producing new heavy elements in the outer core regions.

When the shock wave reaches the outer layers of the star, the visible surface is pushed outward at high speed while being heated up. Thus, the star quickly becomes hotter and vastly bigger than it was; the increased surface area and higher temperature is the cause of its suddenly increased brightness.   As the gas expands it cools and becomes more transparent; as it becomes more transparent we can see deeper into it, and thus are always seeing into a relatively hot layer.   This keeps the supernova at roughly constant brightness for awhile.   Eventually, we begin to see into denser layers which are less transparent, and as they cool off, the brightness begins to fade.   At this time, another source of energy begins to determine the brightness of the star.

The new source of energy is radioactivity, which is the process of energy release as unstable nuclei decay into other nuclei.   In this case, nickel 56 (^{56}Ni) decays into colbalt 56 (^{56}Co), which further decays.   The half–lives of these two nuclei are 6.1 and 78.8 days, respectively, and the decay of ^{56}Co supplies enough energy to determine the rate of decline in brightness.   This is the major source of the brightness of the expanding gas cloud for a year or more after the explosion.

In stars with masses larger than those considered above, the scenario is similar, with one major difference.   In this case, the core probably collapses into a black hole.   At any rate, much work yet remains to be done before we will know how these stars end their lives, and whether black holes are actually formed in the collapse.

The idea that a supernova explosion was involved in the production of a neutron star was proposed for the first time by Zwicky and Baade, in 1933.   This was several years before the physics of the neutron star was worked out by Oppenheimer and his colleagues, and 34 years before the first observational evidence for neutron stars was obtained in the discovery of pulsars.   This was truly inspired astrophysical speculation!

TYPE I SUPERNOVAE

Type I supernovae evidently result from an entirely different mechanism than that which we have just described, because they show no hydrogen in their spectra.   This implies that there is no deep envelope of hydrogen–rich gas to get blown off in the explosion.   The "standard model" will be described here which assumes that the victim is a carbon–oxygen white dwarf in a binary star system.   Further, the model assumes that the triggering mechanism is the accretion of material onto the surface of the victim.   As more material is added, the star approaches the Chandrasekhar limit of about 1.4 solar masses, and the temperature and density in the center increase.

How the material gets dumped on the victim is the center of controversy.   Earlier, a scenario was presented in which a white dwarf in a binary system accretes mass which is dumped onto it by a companion filling its Roche lobe.   The most likely result of this scenario is a nova explosion, but some models have been developed which assume that the right accretion rate would add mass until the Chandrasekhar limit was reached.   It is not clear that the addition of mass from a companion will produce the desired result; apparently it is difficult to avoid a nova explosion.   Further, this scenario describes a parent system which is thought to be too rare to explain the number of supernovae of Type I which are observed.

In any case, if we suppose that the white dwarf is pushed over the Chandrasekhar limit, the result will be the ignition of carbon burning while the white dwarf is still degenerate.   The fact that it is still degenerate is important: if it were not, the energy produced by the nuclear reactions would heat up the star, which would increase the gas pressure.   In a non—generate star, increasing the gas pressure would cause the star to expand, and this expansion would lower the density and slow down the nuclear reaction.   What has just been described is analogous to a thermostat, and it can regulate the rate of the reaction.   Under degeneracy, though, the electron pressure, which dominates, is independent of the temperature.   So, as the reactions heat up the gas, no increase in pressure occurs, no expansion results, and nothing happens to regulate the rate of the reactions.   The result is a thermonuclear runaway. Eventually, the degeneracy is lifted, but by then it is too late.   The material in the white dwarf burns to iron and related heavy elements rapidly, releasing enough energy to blow up the star.   In this mechanism, no neutron star or other stellar remnant is expected to survive.   The star is completely blown up, producing a gaseous remnant.   All hydrogen present before the explosion is burned so that none remains in the remnant.

One of the results of the reaction chain is ^{56}Ni, which supplies energy by radioactive decay just as in the case of Type II supernovae. In this case, the radioactive decay of ^{56}Ni and ^{56}Co is more important than in Type II explosions, and it dominates the light curve.   Models reproduce the observed light curve and spectrum of Type I supernovae well.   The model agrees with the observations in its distribution, too; carbon—oxygen white dwarfs can be present in both Population I and Population II stellar groups, as is observed for Type I supernovae.

SUPERNOVA 1987A

Now that we have discussed generic supernovae, we can discuss some aspects of the subject of this book.   This supernova has been classified as a Type II, because there are hydrogen lines in its spectrum. But its light curve is very peculiar for a Type II (or, even, a Type I, for that matter).   It has not gotten nearly as bright as a typical Type II, and after the initial rise, it has continued to brighten for a couple of months, instead of beginning to drop off.   It is nearly unique in that we have identified the star which blew up using earlier observations, so we know what kind of star it was and how bright it was.   The fact that it is located in the Large Magellanic Cloud has two advantages over what we would have if it were in our own galaxy: we know its distance fairly well and there is little interstellar extinction.   A supernova in our own galaxy might have been brighter if it were not seen through too much interstellar extinction, and that would be an advantage, but we would have a poorer idea of its distance, and it could have considerably more extinction.

In any case, the type of star which has been identified as the victim is surprising: it is a blue supergiant (of type B3). The model of Type II supernovae described above assumes that the star is a red supergiant before it blows up. Why is this important? There are two implications from this difference: the evolutionary state of the core, and the depth of the envelope. Generally, for a star to be on the blue side of the supergiant region would indicate an earlier evolutionary state than is assumed by the model described earlier. On the other hand, the Large Magellanic Cloud generally has a lower abundance of the heavier elements than the Population I stars in our galaxy, and this will tend to make the stars bluer. Further, it is likely that the star has lost mass from its hydrogen—rich envelope sometime during its evolutionary history. This can happen as a result of the mass—loss process described earlier, or it can happen if the star is a member of a binary system. Preliminary models for the light curve of Supernova 1987A fit the observations better if the hydrogen envelope is assumed to be less massive than if no mass—loss had occurred; thus the fact that the star is blue can be accommodated by the theory. The fact that the star is a blue supergiant has the implication that the hydrogen—rich envelope is not nearly as extended as would be true if the same—mass star were a red supergiant. This means that the explosion shock wave will propagate differently than it would through an extended envelope. Preliminary models which take these factors into account seem to fit the major features of the light curve fairly well, including the rather faint early brightness in visible light. As will be discussed later, Supernova 1987A is unusual in that neutrinos were detected for the first time for any supernova explosion. Since the neutrinos were emitted at the time of collapse, they give us an excellent measure of when the explosion occurred relative to the events of the light curve. Since the brightening in visual light occurs only after the shock wave arrives at the surface of the star, we have, for the first time, a determination of the time it takes the shock wave to go from the core to the surface. This was an important part of the information needed to construct the preliminary model.

The light curve of Supernova 1987A is unusual in that it has continued to increase in brightness for a couple of months, before leveling off and decreasing. Some source of energy must cause this brightening, and it must have a different relative importance than in other Type II explosions. In the preliminary models described above, this brightening is assumed to be due to the radioactive decay of ^{56}Ni and ^{56}Co.

CHAPTER 11
SUPERNOVA REMNANTS

PULSARS

In 1967, the radio astronomer Anthony Hewish and his graduate student, Jocelyn Bell Burnell, were using a large radio telescope to search for quasars ("quasi—stellar objects"), at that time a new and exciting phenomenon. They found something else, and their discovery is a nice example of serendipity in astronomy. They were using a technique to find quasars which depends upon the fact that quasars appear, as the name implies, "stellar." That is, quasars are unusual among objects at extra—galactic distances in that they appear star—like—a point source. When seen through our turbulent atmosphere, stars appear to "twinkle," whereas planets do not. Planets are disk—like sources of light, considerably larger than the apparent sizes of stars. Thus, the observation of "twinkling" tells you whether the source is extended or point sources. Point radio sources, seen through the interplanetary medium (mostly hydrogen gas) in our solar system, "twinkle" also, when observed with a radio telescope. Most galaxies, like planets seen through our atmosphere, are large enough that they do not twinkle. Thus, the existence of twinkling would indicate that a radio source probably was a quasar, and not a galaxy. The twinkling of the radio sources can be detected only with radio telescopes with unusually fast response, and Hewish's telescope had been built especially for this use. It was this fast response which made possible their exciting discovery.

The observations were tedious, because they involved not only the recording of the radio signals on long "strip—charts," but the review of the charts afterwards. And, not only did the charts have to be reviewed and measured to find quasars, but the signals had to be separated from noise generated by natural and man—made sources. In October, 1967, Bell noticed a "bit of scruff," which she had seen before, but could not classify. It reappeared each day after an elapsed time of 23 hours 56 minutes, which meant the source was outside the solar system. What was remarkable about the signal was that it was a series of repeating pulses, 1 1/3 seconds apart. She conferred with her advisor, Hewish, but neither he nor other colleagues could suggest what the source was. Repeated pulses are most likely to be man—made, but no man—made source is likely to come and go after elapsed times of 23 hour 56 minutes. At one time, the signal was half—seriously called "LGM—1," for "Little Green Men," but they soon concluded that the source was not artificial.

After a few months, three more *pulsars*, as they are now called, were found. They were remarkable for the incredibly precise repetition of the pulses, and the very short time between pulses. In 1974, Hewish shared the Nobel Prize in physics for his role in the discovery of pulsars and other work. Initially, models which involved the pulsation of very dense objects such as white dwarfs, were proposed. A year before the discovery of pulsars, Franco Pacini of Cornell University published a paper in which he proposed an answer to the puzzle we left at the end of Chapter 4: the source of the energy of the synchrotron radiation from the Crab Nebula. He proposed that a rapidly—rotating neutron star possessing a strong magnetic field would accelerate the electrons. This prophetic speculation was to be vindicated soon, but not to be accepted easily. In 1968, Thomas Gold, also of Cornell University, proposed the general model for pulsars which we accept today: the "lighthouse" model based upon a rotating neutron star. That is, the model proposes that a neutron star is sending out a beam of radiation which is swept past the observer as the star rotates. Each time the beam sweeps past the observer, a pulse is seen. How is the beam of radiation produced? Evidently, it results from the motion of electrons in an intense magnetic field; the tight beam results from a "beaming" effect described by Einstein's theory of relativity and the outward motion of the electrons along the concentrated fields at the magnetic poles.

Why should neutron stars be rotating so rapidly, and why should they have such intense magnetic fields? We know that both rotation and magnetic fields are common in stars. The Sun has a surface magnetic field, and it is rotating with a period of about one month. If you were to compress the Sun down to the size of a neutron star, both the rate of rotation and the strength of the magnetic field would be greatly increased. In the case of the rotation, you can recall the example given in Chapter 6: When a rotating object gets smaller, the rotation gets faster. Compressing the Sun to the size of a neutron star means compressing it by a factor of a hundred thousand. This would reduce the rotation period from a month to about 26 seconds! The real situation for neutron stars is not so simple as compressing a normal star, such as the Sun, down to the dimensions of a neutron star, because the physical processes of evolution and the supernova explosion are very complex, even as they affect rotation and magnetic fields. Nevertheless, this analogy gives a rough idea of the cause of the unusual characteristics of neutron stars.

Initially, Gold's "lighthouse" model of pulsars was not accepted. But a new discovery finally showed that it was the most likely explanation. In 1968, a radio—emitting pulsar was discovered by D. H. Staelin and E. C. Reifenstein of the National Radio Astronomy

Observatory in Green Bank, West Virginia in the Crab Nebula, although they were not able to say precisely which star, if any, was responsible for it.   Its radio output was flickering at a remarkably rapid and constant rate of one flash each 0.033 seconds or about 30 flashes per second.   That made it the most rapid pulsar then known, and made it too rapid to be explained by the pulsation of a white dwarf.

Actually, after only a month of repeated observation, it was obvious that the rate was not constant but was very gradually slowing down. Using a variation of the "turn–the–clock–back" technique, it was possible to deduce that the Crab pulsar could not have existed longer than about 1000 years, which would be consistent with the debut of the Crab Nebula in the great explosion of 1054.   At this point, we have a date computed three different ways and all coinciding within their respective uncertainties: the expansion age, the pulsar slow–down age, and the visual sighting by the ancient Chinese.

When the "lighthouse" model of pulsars was proposed, it was realized that the energy radiated into space would come from the energy of rotation.   This would result in the pulsars slowing down; the observed slowing of the pulses confirmed the model closely.   An unimaginably intense magnetic field a million–million times that of the Earth emanates from its north and south magnetic poles, which lie close to the neutron star's equator.   In the Earth and also in the Sun, by contrast, the magnetic poles are only a few degrees away from the rotational axes and are only about one gauss in intensity.   The thousand–million–gauss magnetic field reaches out, grabs onto the plasma (a volume of space filled with hot gas composed of electrically charged particles) filling up the Crab Nebula, and uses it as a brake to slow down the neutron star.

So far, pulsars had only been observed at radio wavelengths. Could they be seen pulsing optically, as well?   The first optical observation of a pulsar involved, as you might expect, the one in the Crab Nebula.   The discovery was made by three astronomers at the Steward Observatory of the University of Arizona in January, 1969.   The observers, W. J. Cocke, M. J. Disney, and D. J. Taylor, took advantage of the fact that the period of the pulsation was already known from the radio observations.   They used a computer to sum up the pulses over a period of time so that the weak flashes could be seen.   Their work was remarkable, because they used a relatively small telescope, the original Steward Observatory 0.9–meter telescope, located on Kitt Peak.   They could not tell which of the two central stars in the nebula was the source, however.

Their observations were soon confirmed by other observers. Particularly exciting were the observations made in February, 1969, when Joe Miller and Joe Wampler used a device akin to a stroboscope, attached to a TV camera on the 3–meter reflector at Lick Observatory atop Mt. Hamilton in California, to see with their own eyes that it was the southwest star which was flickering.   A strobe was needed because the human eye cannot perceive any flickering which occurs more rapidly than about 20 times per second.

In that same year, 1969, astronomers from the Naval Research Laboratory, in Washington, D. C., and the Columbia University, in New York City, observed the rapid flickering of x—ray radiation from the Crab pulsar.   Because the atmosphere is not transparent to x—rays, they launched rockets carrying instruments capable of detecting the x—ray pulsing; the exact rate coincided precisely with that in the radio and optical parts of the electromagnetic spectrum.

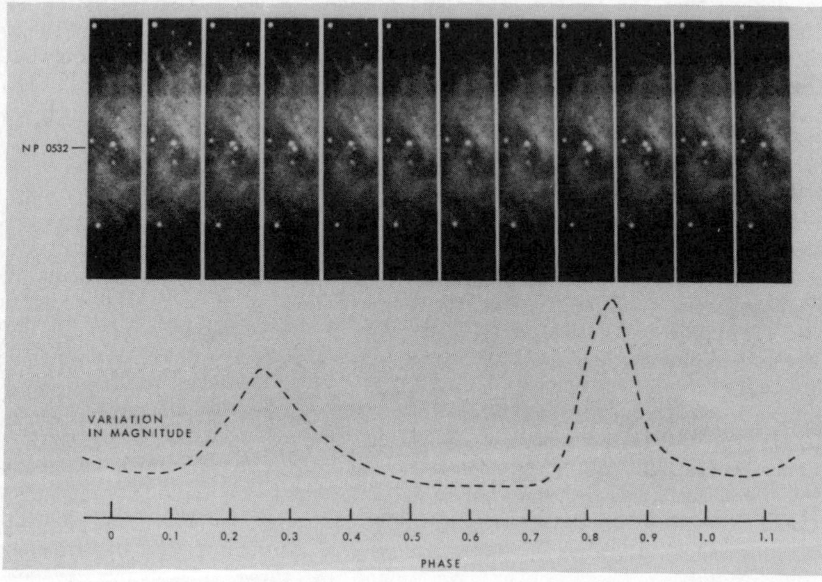

Figure 1.   Photographs of the Crab pulsar in visible light, showing the variation in brightness over a period of one—thirtieth of a second.   The bottom half of the figure is a graph of the light variation.   These photographs were taken on the 2.1—meter telescope at Kitt Peak in November, 1969 (Photo courtesy NOAO).

Since these exciting discoveries, over 400 pulsars have been found. Curiously, only five pulsars have been found in supernova remnants. Does that mean that the remainder are not produced by supernovae? Not at all!   We take this to mean that those not associated with supernova remnants have remnants that have become undetectable, either through cooling and dissipation, merging with the interstellar medium, loss of coupling of energy from the pulsar into the nebula, resulting in cooling to invisibility, motion of the pulsar away from the remnant, or some combination of these.   Many of these pulsars are quite old.   Their numbers are consistent with their origin in supernovae at the estimated rate of supernovae in the galaxy.

Supernova remnants can be classified into two types, "shells" and "plerions."   The "shells" appear like rings of gas, whereas "plerions" appear bright over their outer central region.   The "shells" appear to be nothing but a "bubble" of gas blown outward.   The gas cools off with time and eventually will become invisible, although it may continue to radiate in clumps as it slams into clouds in the interstellar medium.   Of the "historical" supernovae, the supernova of 1006, Tycho's Supernova, Kepler's Supernova, and the remnant of the supernova of 1667 (Flamsteed's supernova?) all show shells.   In none of these is there any evidence of a pulsar.   Tycho's and Kepler's supernovae have generally been considered to be Type I supernovae; remember that Type Is are not expected to leave any "stellar" remnant.

The Crab Nebula is an example of a plerion; we have seen that the region of strong synchrotron radiation is evidently powered by the spinning neutron star.   This also explains the phenomenon of "waves" and other brightenings seen in the central regions of the visible nebula.

The remnant of the supernova of 1181 A.D. has been identified with the radio source 3C 58 (a designation meaning it is the 58th entry in the Third Edition of the Cambridge University Catalogue of Radio Sources).   It also is a plerion.   The Crab Nebula is clearly another example of a plerion.   There is a small, hot star (a "compact object") in the center of 3C 58 but, curiously, all attempts to prove it is a pulsar have failed.   This is not alarming because, if the observer is not located along the line on the sky over which the beam sweeps, the pulses won't be seen.   We estimate that perhaps 90% of the pulsars aren't seen because the beam misses us.

There are other cases like the Crab Nebula in which both a plerion and a pulsar are seen.   One case is the Vela supernova remnant (see next section); it has the distinction of being the closest remnant to the Earth, only about 1600 light–years away.   The pulsar is surrounded by a small cloud which is emitting synchrotron radiation, but it is much fainter than the Crab Nebula.   An example in the Large Magellanic Cloud, known as 0540–69.3, is strikingly similar to the Crab Nebula.   The pulsar flashes 20 times a second, and the surrounding nebula radiates with synchrotron emission.   It is, perhaps, unfortunate that it is so far way, because detailed observations are made impossible by its faintness and small apparent size.

NEBULAR SUPERNOVA REMNANTS

The sky is littered with the remains of stars which perished in supernova explosions long ago.   Their gaseous innards, flung outward at speeds of thousands of kilometers per second now appear as fuzzy patches of light, roughly circular in form.   We now know the Crab as only the most conspicuous of the *supernova remnants* in the sky.

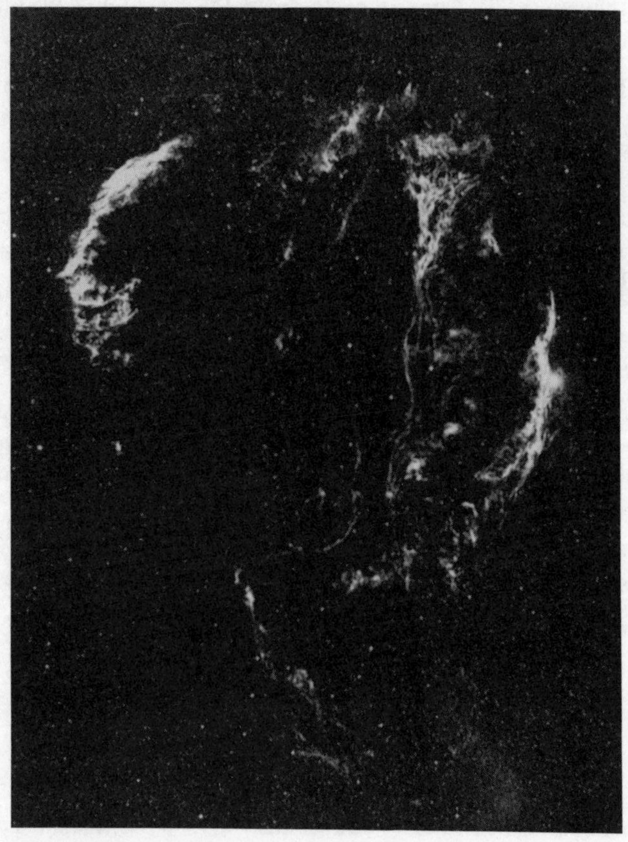

Figure 2.   The Cygnus Loop, a very old supernova remnant, seen at visible wavelengths (Courtesy Palomar Observatory).

To date, astronomers have detected 135 supernova remnants. The techniques used to locate them involve visual observation (looking through a telescope and seeing the characteristic fuzzy patches of light), photography (using time exposures to bring out nebulosity too faint to see by visual inspection), radio astronomy (taking advantage of the fact that most supernova remnants emit radio waves), and x—ray astronomy (taking advantage of the fact that some remnants emit x—rays also). As we have seen, it is possible to date a supernova remnant by using its expansion rate and its present angular size in the sky and then "turning—back—the—clock" to estimate when the nebula would have been a point, that being the moment when the explosion originally occurred. In this fashion it was found that most of the known supernova remnants resulted from explosions which occurred in prehistoric times. The oldest known remnant, IC 443, occurred 60,000 years ago.

A beautiful example of an older supernova remnant is the Cygnus Loop (see Figure 2), which includes the Veil Nebula (NGC 6960, 6992, 6995 and 6979). The Veil Nebula can be seen in small telescopes and binoculars, and is part of a circular ring about 2 1/2 degrees across (about five times the apparent size of the Moon). It is the remnant of a supernova which exploded a few tens of thousands of years ago. Probably most of the emission from the Veil Nebula results from interaction with the interstellar medium.

To our knowledge, the Gum nebula and the Vela supernova remnants are the remains of the closest supernova explosion. We see the remnant on photographs, as the strong x—ray source Vela X—1, and as a radio pulsar flashing 11 times per second. It does not appear in Table I with the historical supernovae because the explosion which produced it was not recorded. Various estimates of the age (of both the supernova remnant and the pulsar itself) place the outburst somewhere around 8000 B.C. Its distance has been estimated at only 1600 light years from Earth.

The Gum Nebula is unique among known supernova remnants. It appears in the southern sky as an enormous luminous wreath approximately 60 degrees in diameter, completely surrounding the smaller Vela supernova remnant and having the Vela pulsar at its center. Its presence was not obvious until Colin S. Gum, a South African astronomer, used special filters and long time exposures to take wide angle photographs of the sky. Colin Gum died at the relatively early age of 36, killed in a 1957 skiing accident only a few years after finding the remarkable nebula which still bears his name. From the fact that the Vela pulsar is only 1600 light years away from Earth and the fact that the nebula has a diameter of 60 degrees, it can be computed that the leading edge of the Gum Nebula is only 800 light years from the Earth, which is half way between us and the center of the original explosion. There is, however, nothing to worry about in the very near future. The Gum Nebula is probably not expanding, although the smaller (and more distant) Vela supernova remnant definitely is.

Here is what probably happened. When the star blazed so brightly 10,000 years ago, it was as if a flash bulb had gone off in the galaxy. As the light streamed throughout the galaxy in all directions, it caused atoms of the interstellar gas to become excited and to fluoresce. The flood of light itself has already passed beyond and left the galactic plane (at the speed of light, 300,000 kilometers per second, that takes only one or two thousand years), but large regions of gas (outlining a generally circular area of the sky) are still shining as the atoms gradually return to their normal unexcited state. This is the Gum Nebula. The hot gas from the expanding blast wave of the Vela supernova remnant is another story. That is heading our way and will impact on the Earth; but, travelling at 1000 kilometers per second, that will not happen for another 500,000 years.

CHAPTER 12
THE INTERNATIONAL
ULTRAVIOLET EXPLORER

Chapter 1 told of the discovery of Supernova 1987A. Chapters 2 through 11 dealt with the background and theory of supernovae in general. We now pick up the story of Supernova 1987A.

With the supernova discovered and the word sent out, astronomers began observations as quickly as possible. Fast response was most important for observations in the ultraviolet. It was expected that the brightness of the supernova would drop fastest in the ultraviolet, and if ultraviolet spectra of the supernova were to be obtained, they needed to be obtained quickly.

THE INTERNATIONAL ULTRAVIOLET EXPLORER

The International Ultraviolet Explorer (IUE) was launched in January, 1978, from Cape Kennedy. The IUE consists of a 0.45–meter (18–inch) f/15.0 Cassegrain telescope, two spectrographs, a photometer (which does double duty by helping out in pointing), and support equipment such as power and communications.

Designed to last three years, and really only expected by some to last a single year, the IUE is now in its 10th year of operation. The IUE is shared between NASA and the European Space Agency (ESA), with NASA having two 8–hour shifts per day and ESA having one. In a synchronous orbit high above the Atlantic ocean, communication with the IUE is maintained from the NASA tracking station at Wallops Island, Virginia, and from the ESA tracking station in Villafranca del Castillo, Spain.

Figure 1.   The International Ultraviolet Explorer (IUE) in its geosynchronous orbit high above the Atlantic Ocean.   The large solar cell panels sticking out the side are now in their 10th year of supplying power to the IUE.   The long "snout" helps keep stray light from the highly sensitive telescope.   The constellation of Orion (the Hunter) can be seen in the background (NASA artist's drawing).

At the engineering control center at NASA's Goddard Space Flight Center in Greenbelt Maryland, just off of the beltway around Washington, engineers keep a constant watch on the health of the IUE. In a nearby building is the telescope control center, so familiar to many astronomers. Encompassing a small glassed—in area just off the hall, the control center consists of a maze of displays and keyboards. Astronomers are assisted by staff operators, such as Holly Abraham (see figure 2), and Resident Astronomers, such as George Sonneborn and Rich Arquilla.

Figure 2.   Some of the displays and keyboards at the astronomer's control center.   Holly Abraham is at the controls (Photo taken by Russell Genet).

ULTRAVIOLET OBSERVATIONS OF SUPERNOVA 1987A FROM SPACE

It was Tuesday morning, February 24, 1987, and Robert P. Kirshner was a bit late in getting to work at the Harvard—Smithsonian Center for Astrophysics, in Cambridge, Massachusetts. He started to clear off his cluttered desk when the phone rang. It was Craig Wheeler, calling from his corner office on the 16th floor of Moore Hall at the University of Texas.

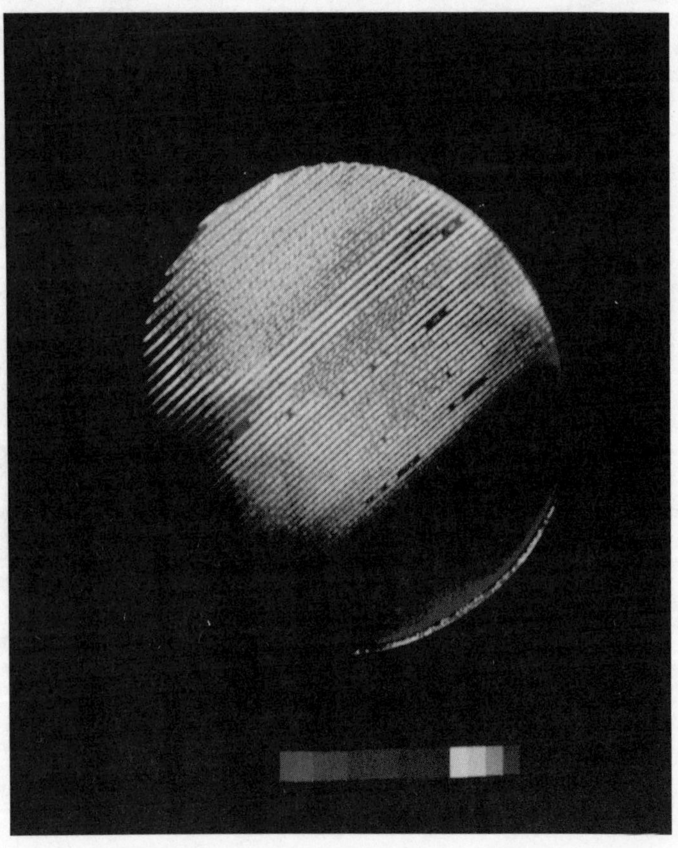

Figure 3.   A spectrum of Supernova 1987A received from the IUE.   The spectrum is spread out similarly to lines of print in a book.   If you look closely at the spectrum you can see a number of double absorption lines. These are from gas clouds in the LMC and in our own galaxy. They are displaced from each other because of the large difference in velocity (the Doppler effect).   A sharper "electronic eye" can see absorption features from many different clouds moving at various velocities.   (Courtesy of NASA's Goddard Space Flight Center.)

"Have you heard about the naked—eye supernova sighted by Ian Shelton, an employee of Toronto University at their southern observatory?" Wheeler asked in a casual, low—key voice.

Kirshner recalled the prank that had been played on him while he was attending a supernovae meeting overseas. Wheeler and a few others had made up a fake telex that read:

DR. ROBERT P. KIRSHNER: BRIGHT SUPERNOVA FOUND IN M—31. RETURN IMMEDIATELY.

Not to be taken in again, Kirshner replied in a relaxed manner, "No he hadn't, but thanks for the information Craig."

Could it be true? Kirshner knew how to find out, and walked over to Brian Marsden's office. Marsden was on his terminal, he was talking on one phone, and another one was ringing. The telex was typing away. Kirshner instantly knew this madhouse could only be the result of a supernova.

Kirshner had, with considerable foresight and a bit of luck, put in a proposal to NASA that if there were a supernova brighter than 12th magnitude that he was to get special time on the International Ultraviolet Explorer (IUE) to observe it. This one was certainly brighter than 12th magnitude! Would NASA come through?

He walked back to his office and looked up Yoji Kondo's phone number. Kondo is the NASA Project Scientist in charge of the IUE. Just as he was reaching for the phone it rang. It was Kondo. "Would Bob be interested in observing the just—discovered supernova with the IUE?"

Kirshner was certainly interested! "Could the IUE staff get the observations going?"

George Sonneborn, the IUE Resident Astronomer at the Goddard Space Flight Center, took charge. First, the astronomer using the telescope, Gary A. Wegner, from Dartmouth College, was asked to yield the last part of his shift. He was glad to do so when he heard about the supernova. Exposure times were calculated, and the IUE was slewed towards the supernova. By 2:00 PM EST, ultraviolet photons from the supernova in the LMC were being recorded for the first ultraviolet spectrogram.

During the first week of the supernova, about 80% of the IUE time was spent observing the supernova. Besides Bob Kirshner and George Sonneborn, Michael Crenshaw, Ronald Pitts, Richard Arquilla, and Chris Shrader all ended up helping with the observing as the weeks rolled by. Bob Kirshner, realizing that this project was too much even for him, asked George Sonneborn to be an official co—investigator with him. The two of them worked well together, and with the rest of the IUE staff.

While Kirshner had expected the ultraviolet light from the supernovae to fade, it faded even faster then expected. On Tuesday, the first day, it took only a 15 second exposure to obtain a spectra, while by Friday the exposures took 15 minutes.

Figure 4.   Yoji Kondo, NASA's Project Scientist for the International Ultraviolet Explorer.   As the "boss" for NASA's most successful space telescope, Yoji has worked with hundreds of astronomers for almost a decade, helping them to bring in the rich harvest of this remarkable instrument. (NASA photo.)

Pre—discovery photographs had shown two stars at the position of the supernova. One was Sanduleak −69° 202. This was the 202nd star at declination −69° in the catalog of Nicholas Sanduleak, an astronomer presently at Case Western Reserve University. Sanduleak made the observations for his catalog while at Cerro Tololo Interamerican Observatory. Sanduleak −69° 202 was a normal−looking giant star of type B3 I, a hot fairly massive star. At 12th magnitude, it did not appear very bright from earth. Near it was an even fainter star, type unknown.

While it was unheard of for a hot "early" type star to turn into a supernova, everyone had, by Monday, pretty much assumed that Sanduleak −69° 202 must be the progenetor—the star that exploded. Nothing else seemed possible. This meant that when the ultraviolet radiation from the supernova died down, the light from the other, fainter single star would come through.

By Monday, the ultraviolet light from the supernova was almost entirely gone and a 90 minute exposure showed two background stars. Bob and George were shocked to see light from **two** stars instead of one. What was going on here? Photographs taken before the supernova had exploded had shown two stars. One had exploded. Simple mathematical calculation suggested that there should only be one left!

Kirshner and Sonneborn concluded that the two stars showing in the ultraviolet must be Sanduleak −69° 202 and the faint companion, and the supernova progenitor must have been something else—something not visible on the pre−discovery plates. This would be most unfortunate indeed. Everyone had been pleased that the Sanduleak star had blown up, as then there would not be any close−by relatively bright star to interfere with sensitive measurements. Now it appeared that 12th magnitude Sanduleak −69° 202 was going to ruin the game. What a cruel thing for nature to do. With the whole sky to have stars in, putting one literally on top of the supernova seemed unfair. What to do?

Robert Kirshner has lots of friends in the astronomical community, always willing to help him out. Soon he received the following letter:

MASTER BLASTER STAR REMOVAL
182 Stellar Drive
Crater Lake, Oregon 97626

March 11, 1987

Dr. Robert Kirshner
Astronomy Department
Harvard University
Cambridge, Massachusetts 02138

Dear Dr. Kirshner;

We at Master Blaster Star Removal Company recently read in the Boston Globe of your despair. It seems that a bright star is in the way of your astronomical observations. This problem has been encountered before. We know that yours is an important discovery, and should be studied without obstruction.

We may be able to fix your problem. The Master Blaster Star Removal experts have devised a system never dreamed of before. Our system has proven to be on target, 100% accurate, and completely successful. You astronomers have never found the 10th planet. You never will. We found it, and used it to test our system.

Our crew has successfully destroyed many planets, stars, asteroids, and moons, and there is a long waiting list, so don't delay! Contact us immediately! The price is reasonable, but we expect prompt payment for the transportation of our Blaster Crew.

Remember our motto; "We've got class; We'll blow that mass!" We accept Visa or Blastercard.

Sincerely,

Alistar Nukitall Antimatter

The word was out: "Kirshner says supernova progenitor not Sanduleak −69° 202."

Meanwhile, pre−discovery photographs were being looked at very closely again.   Richard M. West at European Southern Observatory thought there was a hint of a third star.   The supernova was too bright in   the visible wavelengths to take another photograph, but perhaps observations made earlier by the IUE itself could be reanalyzed to solve the puzzle.   Some spatial information along one axis could be squeezed out of the IUE data if the slit of the spectrograph happened to (accidentally) lay in the right direction.   It did in the early data, and Bob and George were able to determine that the second star was not quite at the location of Sanduleak −69° 202.   Thus the Sanduleak star was the supernova after all!

And what did astronomers now think of Bob's first proposal that the supernova was not Sanduleak −69° 202?   Case Western Reserve University astronomer Peter Pesch thought Bob's idea was a red herring.

Figure 5.   Robert Kirshner holding can of "Herring Fillets in Tomato Sauce" received from Peter Pesch (Photo taken by Russell Genet).

The result of all the IUE observations was that important observational constraints were placed on the theorists. Models they came up with had to fit the IUE ultraviolet spectroscopic data and the unexpectedly rapid decline in the ultraviolet.

The data were also useful in other ways, such as analysis of the dust and gas between the supernova and the earth. As the light from the supernova passed through gas and dust clouds, dark absorption lines were introduced into the spectra. From the exact wavelengths of the absorptions, Harvard–Smithsonian Center for Astrophysics scientist Andrea K. Dupree was able to draw conclusions about the nature of the interstellar medium, and in this case, the intergalactic medium.

Figure 6. Andrea K. Dupree, Harvard–Smithsonian Center for Astrophysics expert on stellar physics and the interstellar medium. She analyzed the impact of the interstellar medium on the ultraviolet light received from the supernova by the International Ultraviolet Explorer. (Photo by Russell Genet.)

ULTRAVIOLET EXPLORERS

ROBERT P. KIRSHNER

Robert P. Kirshner was born in 1949, and lived in Rome, New York, before settling down in Sudbury, Massachusetts, where his parents still live. As a young boy, Bob inherited a basement—full of electronic parts which he used to build a Tesla coil, oscilloscope, and other scientific paraphernalia.

As with other kids of his age of a scientific bent, Bob scoured the public library for books on astronomy and other sciences, reading those by Fred Hoyle, George Gamow, and others. A reading of Thompson's "Calculus Made Easy" taught him this important mathematical subject well before college.

Bob became interested in observational astronomy while still in high school and was selected to attend a special National Science Foundation program for potential future scientists. Bob spent a summer at Ohio University in Athens, Ohio, where he and another High School student and future astronomer, William Herbst, were turned loose on the 0.22—meter refractor on top of the Physics Department building. While this did not result immediately in science of journal quality, Bob and Bill had a good time exploring the heavens, and both became professional observational astronomers.

Bob entered Harvard University as an undergraduate in 1966. A freshman seminar in astronomy conducted by Allen Maxwell was fascinating and Bob determined he would become an astronomer. The combination of inspiring instructors and classmates added to the rising excitement in the astronomical community over the recently discovered quasars and pulsars—the latter discovered while Bob was still an undergraduate at Harvard.

Upon graduation from Harvard in 1970, Bob went to graduate school at the California Institute of Technology on a National Science Foundation fellowship. Assigned to work with the well—known Cal Tech/Mt. Palomar astronomer Bev Oke, he was asked, "Well young man, what do you want to work on?" Without hesitation, Bob told the eminent Oke, "Supernovae."

As luck would have it, a supernovae in M—101 had just flashed into human awareness, and Bev Oke handed Bob a pile of spectroscopic observations to analyze. Bob was invited to present his analysis of the supernovae M—101 at an international meeting held in Tucson in 1971. Bob learned at this meeting how hazardous supernovae research could be. While walking down a Tucson side street, he, along with fellow researchers Craig Wheeler and Jerry Ostriker, was mugged. More than the M—101 Supernovae, it was this aspect of Bob's research that first brought him to the attention of the astronomical community! Along with a reputation for keeping first—rate supernovae company, Bob received a cast for his arm—broken in the struggle with the mugger.

With broken arm, but with intact spirit, Bob's next stroke of luck arrived in the form of a supernova in NGC 5253. Reaching 8th magnitude, this was the brightest supernova since 1937. The other half of Bob's most recent good fortune was Cal Tech's just completed 1.5—meter telescope on Mt. Palomar. It had been finished so recently that there had not yet been enough time to schedule observations. Bob took care of that—he observed the NGC 5253 supernova every night!

The International Astronomical Union, which meets every three years, held its 1973 General Assembly in Sidney, Australia. Supernovae were a hot topic, and Bev Oke was invited to give one of the special talks. At the last minute he could not go, and young Kirshner filled in for him. Bob was up to the job, and the astronomical community realized that a new supernovae expert had arrived.

Bob's doctoral dissertation was on the spectroscopy of supernovae, and the Astrophysical Journal version became a classic. Accepting a "post—doc" position at Kitt Peak National Observatory, Bob worked with Roger Chevalier on supernovae remnants. Spectroscopy of the knots in Cassiopasia A, a supernova remnant, clearly showed that the knots were strongly enriched with heavy elements. It was gratifying to be able to show that observations of these remnants matched the expectations generated by the theorists and their mathematical models of stellar interiors under catastrophic conditions. Fritz Zwicky, Robert Oppenheimer, Willie Fowler, Stirling Colgate, Stanley Woosley, and others were correct!

After his postdoc was completed at Kitt Peak National Observatory in 1976, Bob accepted a position at the University of Michigan, which had a strong astronomy department under the leadership of W. A. Hiltner. The University of Michigan had just moved their 1.3—meter telescope out to Kitt Peak for clearer skies. It wasn't heavily scheduled yet, would young Prof. Kirshner be interested in some observing time? With a couple of eager grad students in tow, Bob headed back for more Arizona observing—supernovae remnants, of course—whenever he got a chance. Clear, warm Arizona was a welcome relief in winter from the cold and clouds of Michigan. When Al Hiltner retired, Bob took over as chairman of the Astronomy Department.

In 1984, Bob accepted a joint position as a Professor of Astronomy at Harvard University and a Research Astronomer at the Harvard—Smithsonian Center for Astrophysics (known as the CfA). One of his first actions at CfA was to send a proposal to Yoji Kondo at NASA for making observations on the International Ultraviolet Explorer (IUE) of any supernovae bright enough to be within the IUE capabilities. Several were just bright enough, and Bob found himself traveling to NASA's Goddard Space Flight Center to make the observations. In renewing his proposal to NASA, Bob with great foresight put in a request for extra time "if a supernovae should be brighter than 12th magnitude." The light from Supernovae 1987A was less than a year from earth when Bob thoughtfully added this "clause" to his proposal which was approved by NASA.

Figure 7.   Robert P. Kirshner in his office at the Harvard—Smithsonian Center for Astrophysics. (Photo taken by Russell Genet.)

GEORGE SONNEBORN

Born in St. Paul, Minnesota, George Sonneborn soon moved to Sidney, Ohio, a small community not far from Dayton, Ohio. At age 7, George was given, as a present, the Golden Book On Astronomy. George attended the Walnut Hills High School in Cincinnati, Ohio, where he pursued his interest in natural history, and was able to take an advanced course in physics and astronomy.

In 1969, George began his undergraduate studies at Ohio State University. George excelled at mathematics and physics, but it was a two—quarter sequence in Astronomy for Science Majors that really absorbed his interest. Ohio State has a large astronomy department with a long tradition, and George Collins put young Sonneborn to work on modeling pulsational instabilities in rapidly rotating massive (B type) stars. George received his Ph.D. in 1980, taught Physics at OSU for a year, and then in 1982 accepted a position with Computer Sciences

Corporation (CSC) as part of the International Ultraviolet Explorer (IUE) team.

For his first two years on the IUE, George scheduled astronomers and observations. He was then involved with instrument monitoring and calibration—measuring changes in the sensitivity of the on—board sensors. For the last two years he has been in charge of telescope operations.

Figure 8. George Sonneborn in his office at NASA's Goddard Space Flight Center, Greenbelt, Maryland. George's office is just down the hall from IUE operations (Photo taken by Russell Genet).

SUPERNOVA 1987A AND THE INTERSTELLAR MEDIUM

Scientists studying the interstellar medium consider the supernova to be simply a light bulb that shines exceptionally brightly behind all of the material along the line of sight. By taking advantage of this "light bulb," Andrea Dupree was looking for several things. First the material in our galaxy and especially the halo of our galaxy could be detected by the signatures of highly ionized lines—lines such as C IV (carbon with three electrons stripped away) and Si IV (silicon with three electrons also stripped away). Secondly, the line of sight of the supernova passes through gas in the outer parts of the large Magellanic Cloud. Could a halo be detected there too? Would it be different from the halo around our galaxy? And thirdly, could she find any hints of the damage that the supernova did to the interstellar medium? The rapidly expanding supernova was expected to give out a strong flash of ionizing radiation as the shock of the collapsed expansion reached the surface of the star. Atoms and ions exposed to this intense light would be stripped of their electrons. Could this be seen in the ultraviolet spectra?

George Sonneborn had been careful to take several high dispersion spectra of the supernova when it was at its brightest. Andrea Dupree and George Nassiapoulous, a computer specialist at the Harvard—Smithsonian Center for Astrophysics, eagerly awaited the magnetic tapes that Bob Kirshner hand carried back from the IUE Observatory in Maryland. It would take intensive computer reduction to extract the signal from the tapes and then display it on a video screen to search for the absorption features. Nature can be kind and sometimes helps astronomers. Those studying the interstellar material in the Large Magellanic Cloud were lucky that the LMC is moving away from our galaxy at over 300 kilometers per second. Absorption by the gas in the LMC will be separated in wavelength, due to the Doppler shift, from the signature of the same type of gas in our own galaxy. This separation means we can clearly identify the origin of the absorption features.

Andrea could see a pattern of strong absorption features from the important ion C IV—several at the wavelengths corresponding to gas in our galaxy—and another cluster of them from the gas from the LMC. The features from the LMC were so strong, in fact, that they absorbed much of the light from the supernova. She and Nassiapoulous quickly measured their strengths and ran back to an open copy of the Astrophysical Journal where an important first paper on the interstellar medium had been published by Blair Savage, from the University of Wisconsin at Madison, Wisconsin. When the IUE was first launched, Blair realized that the satellite was far more sensitive to faint light than the original specifications had required and, most importantly, long exposures (up to about 24 hours long) could be taken with the satellite. Blair obtained such long exposures towards the brightest stars in the Large Magellanic Cloud, and he and his colleagues did find the absorptions from the halo of our galaxy and from the LMC. But now this supernova was a factor of 100 brighter than the other stars in the LMC. So instead of an exposure that lasted for hours, IUE could capture the light in only a few minutes. Thus the long—sought absorption features could be discovered directly in one spectrum—including many weak lines not seen before!

The numbers in Savage's table told an important story—the lines from the hot ions in the LMC towards Supernova 1987A were stronger—much stronger than towards most other "normal" stars in the LMC, which indicated a great deal of ionized material along the line of sight to the supernova. And there was also a sharp feature in these lines indicative of gas that was moving away from the supernova at a velocity of 70 kilometers per second. Could this be material ejected by Sanduleak

−69° 202?   The pre−supernova star was an evolved star and must have lost some of its material in a stellar wind.   Was this little absorption blip from the fragment of a massive shell ejected by Sanduleak −69° 202?   The answers are not in yet and the astronomers must say "time will tell."   What they mean in this: if the blast of ionizing radiation caused these features and ionized the dense material expelled by Sanduleak −69° 202, then we expect the features should disappear in a few years as the ions return naturally to a less excited state.   And so the high dispersion spectra are being watched carefully!   Each magnetic tape with a high dispersion spectrum is eagerly anticipated and pounced upon when it arrives.   To date there have been no changes, but over the next few years, Supernova 1987A will be a target of IUE and then HST (Hubble Space Telescope) observations to search for the subtle features that give insights into the violent and turbulent gas many hundreds of thousands light years away.

CHAPTER 13

OBSERVATIONS FROM THE SOUTHERN HEMISPHERE

When the word was sent out, astronomers in the southern hemisphere began observations as quickly as possible. In many cases, the supernova was too bright for large telescopes, and astronomers had to devise ways of throwing some of the light away. Generally, telescopes are scheduled for months in advance, and they are scheduled to observe other objects. Should the scheduled astronomers be allowed to observe their original non—supernova object, should they be asked to observe the supernova instead, or should the telescopes be "taken over" and previously scheduled astronomers preempted? Formal or informal decisions had to be made, and they had to be made quickly.

In this chapter, we will discuss observations from ground—based telescopes, starting with observatories in Africa and working our way around the southern hemisphere *via* South America and New Zealand to Australia. There are too many observatories, telescopes, instruments, and astronomers to cover them all, so we will have to "sample" a few here and there, with apologies to the majority which simply could not be included. The intent is to give a feeling of what was done, not a complete and exhaustive accounting.

While a number of major observatories have been included, we have also purposely included a number of small observatories operated by amateur astronomers. In the chapter on the discovery, it was clear that amateur astronomers played an important role in the Supernova 1987A saga, and this was to continue in the following months.

SOUTH AFRICAN ASTRONOMICAL OBSERVATORY

When Brian Warner heard about the supernova, on the morning of February 24th, while visiting the University of Texas, he immediately got in contact with the South African Astronomical Observatory (SAAO). As it would soon be dark there, the first optical—wavelength spectra could be obtained. The skies were clear, as usual, at Sutherland, the main location for the telescopes of the SAAO.

Figure 1. The four South African Astronomical Observatory (SAAO) telescopes at Sutherland. From left to right they are the 1.9–, 1.0–, 0.75–, and 0.5–inch telescopes. (Photo taken by R. M. Catchpole, SAAO.)

J. W. Menzies was able to get the first spectra of Supernova 1987A with the 1.9—meter telescope. At the same time, Hartmut Winkler was using the 0.5—meter reflector to obtain photometry in several colors. The supernova was so bright that neutral density filters had to be added to the spectrograph to throw away much of the light. The remaining light was then dispersed and amplified with an intensifier and detected with a solid—state linear array of photodiodes (a reticon). Results from the first spectra were phoned in to Brian Marsden, *via* Brian Warner in Texas, in time to be included in the first IAU Circular on Supernova 1987A. The first spectra suggested that it might be a Type I supernova, but later observations made it clear that it was a rather strange Type II supernova—perhaps strange enough to eventually deserve a new class.

Figure 2. The Sutherland 1.9—meter telescope was used to obtain the first optical spectra of Supernova 1987A. When it was first brought into operation in Pretoria in 1948, it was the largest telescope in the southern hemisphere. (Photo courtesy of SAAO.)

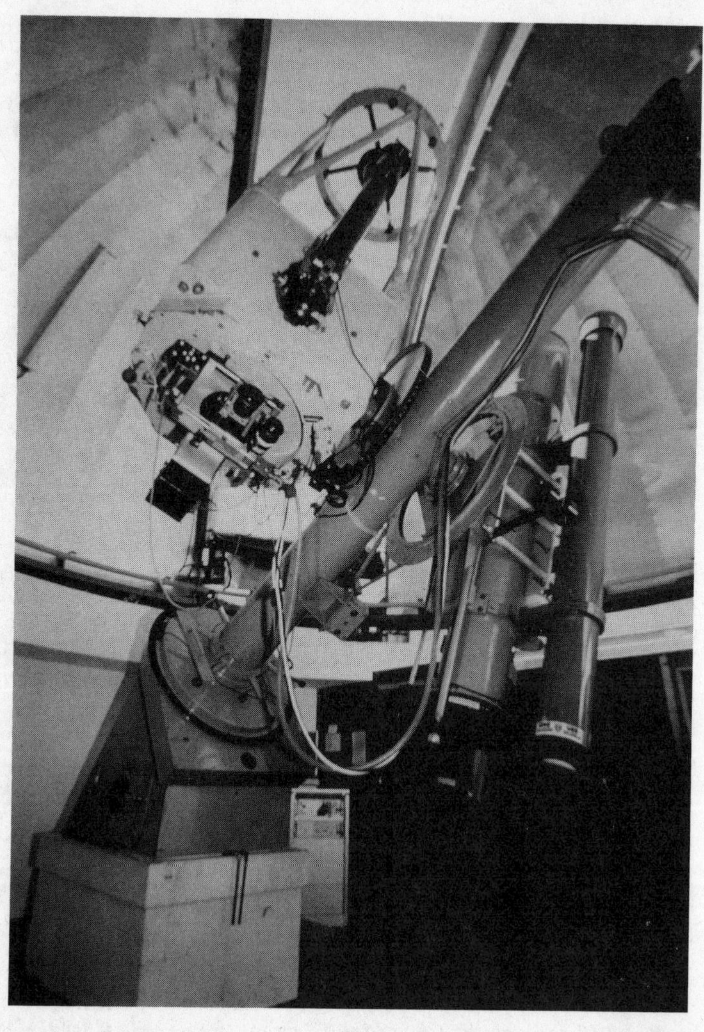

Figure 3   The 0.75−meter reflector used to measure the brightness of the supernova in the infrared.   The Mk II infrared photometer is shown attached to the telescope.   (Photo by Robin Catchpole.)

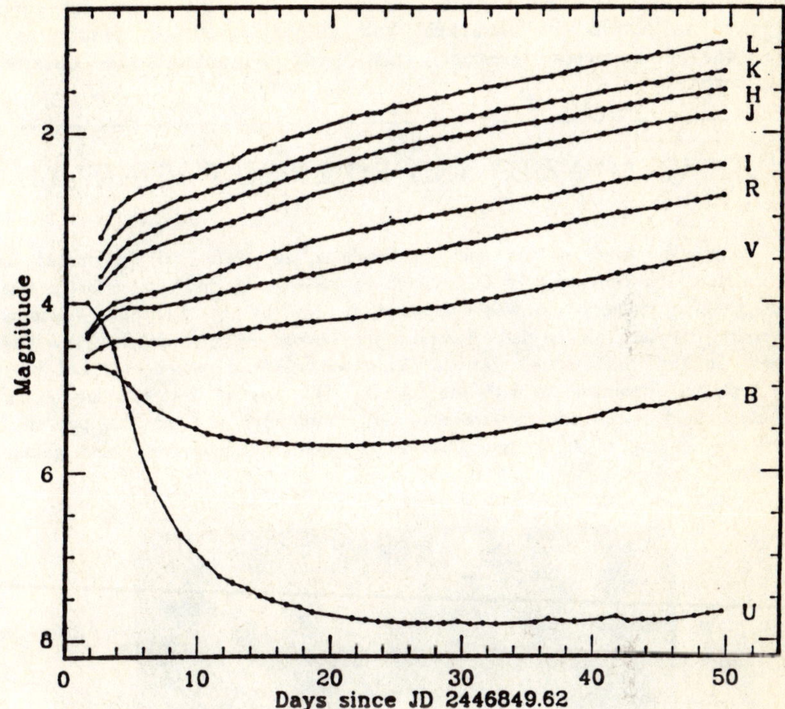

Figure 4.   Photometric results from the 0.5— and 0.75—meter telescopes at the SAAO.   The shortest wavelength ultraviolet "U" band is the curve on the bottom.   Time is plotted across the abscissa going to the right (days since the supernova exploded), while brightness is the ordinate, with brighter being up and fainter being down.   As can be seen, the U (ultraviolet) took a fast plunge.   The "B" band (for blue), dropped, but then came back up. The "V" band (for visual—a yellow band near the peak of the human eye response) mainly was increasing.   The "R" (red), "I" (infrared), and J, H, K, and L bands (further out in the infrared) all increased.   These photometric data from the SAAO and elsewhere placed important constraints on the models.   (Curve from the SAAO Preprint No. 520, to be published by the Monthly Notices of the Royal Astronomical Society.)

Besides obtaining one of the most comprehensive series of optical spectra of Supernova 1987A, the SAAO obtained very accurate photometry of Supernova 1987A.   This photometry measured the brightness almost nightly in a number of different color passbands from the near ultraviolet through the infrared (*UBVRI*; see Figure 4).   The 0.5–meter telescope was used to obtain the ultraviolet and visible wavelength photometry, while the 0.75–meter telescope was used to obtain the infrared photometry.

JOHN W. MENZIES AND ROBIN M. CATCHPOLE

John W. Menzies was born in Brisbane, in 1940.   He graduated in Physics from the University of Queensland in 1963.   His Ph.D. dissertation was about globular clusters and was completed in 1967 at the Australian National University.   After this he left Australia for Oxford as a Radcliffe Travelling fellow.   In 1977 he married Patricia Whitelock, also an astronomer and they both took up appointments with the SAAO.   His research interests are broad. His main field is optical spectroscopy and photometry and he has published papers on, among other things, globular clusters, X–ray sources, and galaxy redshifts.

Robin M. Catchpole, astronomer at the SAAO.   Robin makes many observations in the infrared portion of the spectrum.

Robin M. Catchpole was born in Scotland in 1943, and first became seriously involved in astronomy in 1962, when, after leaving school, he joined the Royal Greenwich Observatory. In 1966, after obtaining a BSc in astronomy at University College London, he went to South Africa to work at the Radcliffe Observatory. This observatory closed in 1974 and he moved, along with the 1.9–meter telescope, to the South African Astronomical Observatory, where he has worked ever since. His astronomical interests include cool star abundances (the topic on which he obtained his Ph.D. in 1981), infrared astronomy, and infrared mapping on the galactic center.

NIGEL OBSERVATORY

The Nigel Observatory is located on the south–western edge of the town of Nigel, in the Transvaal region of South Africa. It is located in the back yard of amateur astronomer Luciano Pazzi.

Figure 5.   Shown above is the Nigel Observatory building.   A 3.9–meter diameter dome rotates around the bottom structure (Photo by L. Pazzi).

While Luciano makes visual, photographic, and photoelectric observations of many different types of astronomical objects, his specialty is making precise photoelectric measurements of brightness variations of variable stars.

Some mathematical analysis needs to be done with the raw measurements to put them in a form usable by other astronomers. This used do be done laboriously with hand calculators, but today this task is computerized at many observatories, including the Nigel Observatory.

Of course, Luciano could not resist making some measurements of the brightness of Supernova 1987A. These were made along with regular measurements on many variable stars.

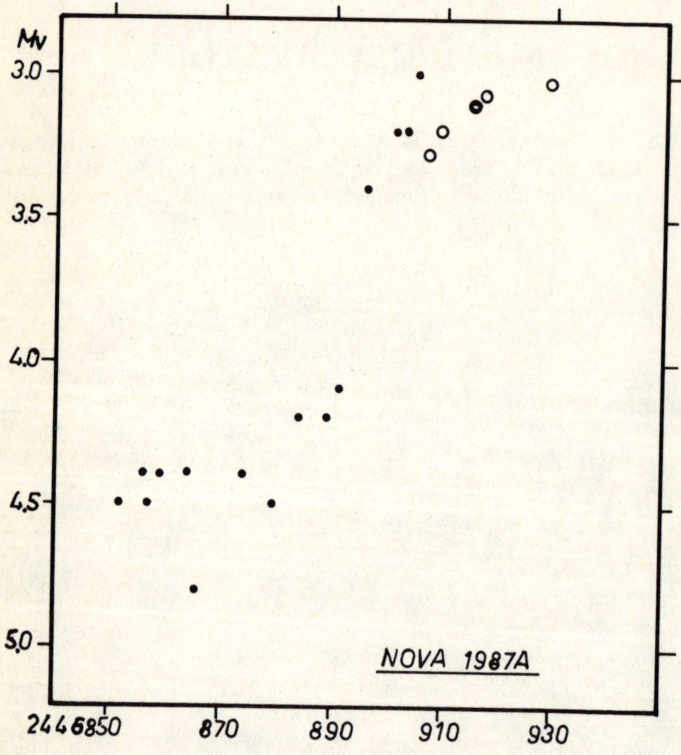

Figure 6. Visual and photoelectric measurements made of Supernova 1987A at the Nigel Observatory. The ordinate is brightness increasing upward. The abscissa is (Julian) days, increasing to the right. The closed circles are visual estimates, while the open circles are photoelectric measurements in the "V" band.

LUCIANO PAZZI

Luciano Pazzi was born in a small town near Bologna, Italy, in 1943. At the age of eighteen, he emigrated to South Africa.

Luciano's love of Astronomy goes back as far as he can remember. At the age of ten, he was observing the moon with a home—made refractor assembled with spectacle lenses. It was only later in South Africa that the serious side of his interest in astronomy emerged. Timing of lunar occultations have been made by Luciano with a portable 0.1—meter Newtonian since 1968. He has participated in many expeditions to observe lunar grazing occultations.

CERRO TOLOLO INTER—AMERICAN OBSERVATORY

Located on the Southern edge of Atacama desert in north—central Chile, Cerro Tololo Inter—American Observatory (CTIO) is part of the National Optical Astronomy Observatories (NOAO), and is funded by the National Science Foundation. It supports about 200 visiting astronomers each year from about 50 different institutions. Observing time is assigned by a Telescope Allocation Committee (TAC) and is based on the scientific merit of the proposals.

The 4.0—meter telescope is the largest telescope in the southern hemisphere, surpassing the Anglo Australian Telescope (AAT) by a small amount. The 4.0—meter has held the title of "Largest In the South" for several years, but plans are afoot for much larger telescopes at the Las Campanas Observatory, and the European Southern Observatory (ESO). The CTIO 4.0—meter telescope is the twin of the Kitt Peak National Observatory (KPNO) telescope near Tucson, Arizona. This telescope was used by Peter Nisenson in conjunction with a speckle camera made at the Harvard—Smithsonian Center for Astrophysics to discover the mysterious companion to Supernova 1987A (see Chapter 16).

CTIO has done well by 1987A, making crucial photometric and spectroscopic observations, and providing the telescope for the discovery of the mystery object. CTIO has a closely—knit, hard—working staff, and they rose to the challenge of 1987A with flying colors.

Figure 7.   The 1.5−meter telescope is Cerro Tololo Inter−American Observatory's second largest telescope. It is equipped for infrared studies, direct photography, photometry, and spectroscopy. Mark Phillips used this telescope for taking spectra of Supernova 1987A. (Photo courtesy of the National Optical Astronomy Observatories.)

Figure 8.   The telescopes on top of Cerro Tololo (Tololo mountain).   The large dome in the rear is the 4.0—meter telescope used by Peter Nisenson and his speckle camera to discover the "mystery object" (see Chapter 16). To the right of the 4.0—meter dome is the 1.5—meter telescope used by Mark Phillips to make spectroscopic measurements of Supernova 1987A.   In front of the 1.5—meter telescope is the 0.9—meter telescope.   Close together on the left are the 1.0—meter telescope on long—term loan from Yale University, and the 0.6—meter Schmidt camera of the University of Michigan.   In the front are the two small domes for the two 0.4—meter telescopes.   These had been shut down, but one was reactivated for photometry of Supernova 1987A (National Optical Astronomy Observatories photo).

Figure 9.   The Cerro Tololo Inter—American Observatory 4.0—meter telescope is the largest optical telescope south of the equator.   Located in Chile, the 375—ton instrument is a twin to the Mayall Telescope at Kitt Peak, Arizona.   (Photo courtesy of the National Optical Astronomy Observatories.)

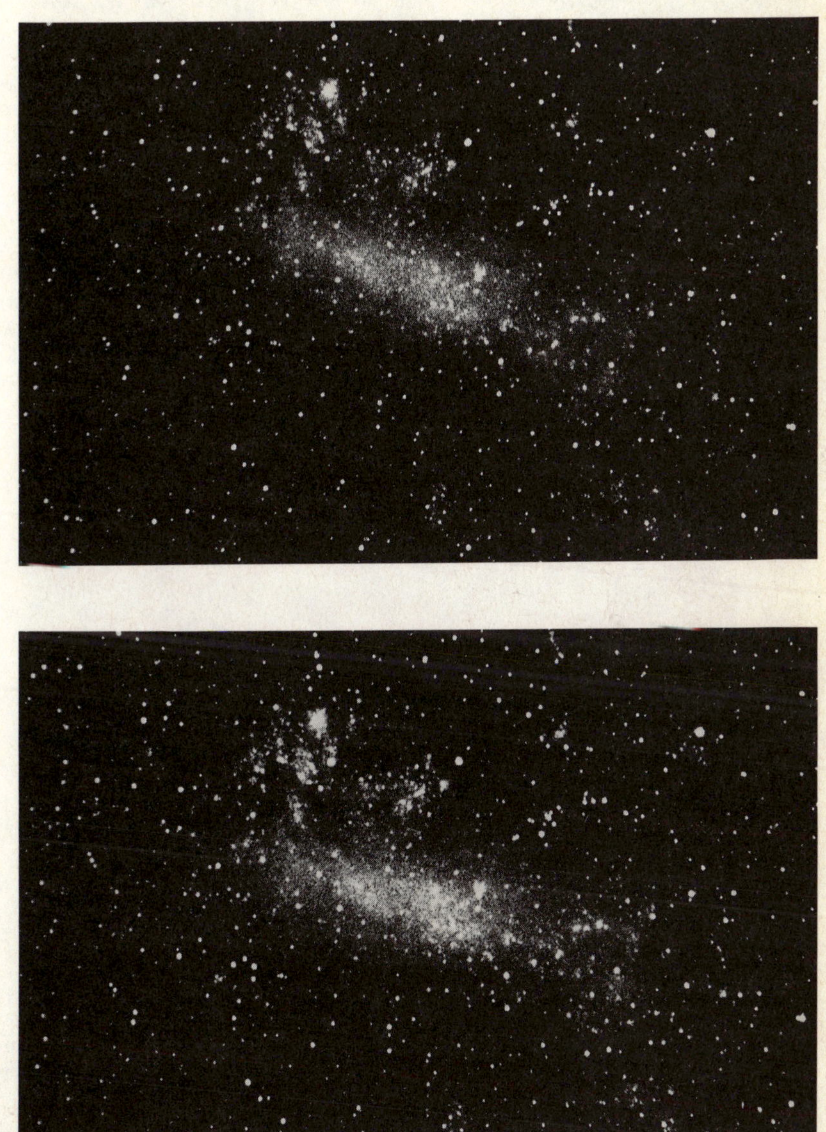

Figure 10 (previous page). A picture of the Large Magellanic Cloud before and after Supernova 1987A, as seen from CTIO. The before picture was taken by Victor Blanco, the long—time director of CTIO (now directed by Robert E. Williams), while the after picture was taken by W. Roberts. (Photos courtesy of the National Optical Astronomy Observatories.)

MARK M. PHILLIPS

Born in San Diego in 1951, Mark had an interest in science at an early age. His older sister, Penny, took a course in astronomy at San Diego State University, and Mark helped her with some of her homework. Astronomy looked like fun, so Mark decided to take the course himself. He liked it so well he decided to become an astronomer, and switched from a math major to an astronomy major. Mark liked observing, and he got to do photometry on the 0.4—meter telescope on Mt. Laguna.

Mark M. Phillips, astronomer at Cerro Tololo Interamerican Observatory. Mark made some of the first spectroscopic measurements of Supernova 1987A. (Photo taken by Russell Genet.)

After receiving his Bachelor's degree in Astronomy from San Diego State University in 1973, Mark went to graduate school at the University of California at Santa Cruz—the location of the headquarters of Lick Observatory. At Santa Cruz, Mark worked with Don Osterbrock, a very experienced spectroscopist.

With the 3—meter (120—inch) telescope at Lick Observatory on Mt. Hamilton, Mark made spectroscopic observations of Seyfert Galaxies—galaxies with active nuclei.

After receiving his PhD in 1977, Mark went to CTIO on a post—doc for a couple of years, and then to a staff position at the Anglo Australian Telescope (AAT) for two years.   When a staff position opened up at CTIO, Mark went back, and he has been there ever since.

ANGLO—AUSTRALIAN OBSERVATORY

The Anglo—Australian Observatory (AAO), as its name implies is a cooperative venture between the British and the Australians.   The main telescope at the observatory is 3.9—meters in aperture, highly instrumented, and very modern.   Only CTIO has a larger telescope in the southern hemisphere, and at 4.0 meters it is for all intents and purposes the same size.   The 3.9—meter telescope is usually called the AAT, an abbreviation for the Anglo—Australian Telescope.

The AAT is located on Siding Spring Mountain in New South Wales, and overlooks Warrumbungle National Park.   It was dedicated in 1974 by Prince Charles, and began regular operation in 1975.

In a number of areas, the AAT has some of the best instrumentation in the world.   This includes outstanding infrared instrumentation that has been used by astronomer David Allen to very good effect over the years.   David, who is the editor of the AAO Newsletter, nicely described the impact of Supernova 1987A on the Anglo—Australian Observatory in his editorial in the April 1987 issue of the newsletter:

EDITORIAL

Shortly before a lovely Sydney sunset on 23 February, about 109 neutrinos passed right through every inhabitant of this planet.   These neutrinos had been journeying for 170,000 years to bring the news that a supernova had erupted in the LMC.   Due to circumstances beyond our control at AAO, we did not detect the neutrinos, so that more than 24 hours were to pass before we knew of the discovery.   From the moment we heard, life has not been the same.   It will be difficult for readers at northern institutions to appreciate the impact of the supernova.   We have shuffled the schedule, taken over service time, overridden observers, we have used up to three instruments per night, even arranging major changes part way through the night; we have cobbled together an instrumental configuration never before used, and we are building an ultra-high resolution system for the coude

focus to take advantage of the SN to probe the
interstellar medium. None of us being supernova pundits,
we have agonized for many man days to make the many
decisions embodied in this list of activities, and we
have done so under a barrage of public and media interest
exceeding in its intensity the height of Halleymania.

David Allen

The AAT has been active in many important aspects of observing
Supernova 1987A. It has played a vital role in the identification of the
"mystery object" (see Chapter 16).

So far, we have only considered telescopes in space and on the
surface of the earth. What about underground telescopes?

CHAPTER 14

THE BIRTH OF NEUTRINO ASTRONOMY

At 7:35:41.37 (Universal Time) on February 23, 1987, a new field of scientific investigation was born: neutrino astronomy. Let us say quickly that neutrinos from beyond the Earth had been detected earlier, in the 1970's, but they were from our Sun, scarcely more than 8 light minutes away from Earth. The first neutrino from the 1987A supernova registered by the exotic detector in a salt mine underneath Lake Erie was the first neutrino ever detected which had its origin outside the solar system and, ironically, it came from outside our galaxy as well.

Neutrino astronomy is fundamentally new, like say, the start of radio astronomy in the 1930's (when Carl Jansky first picked up static from the center of our galaxy) or x—ray astronomy (when American astronomers launched captured German V—2 rockets to study the Sun from above the Earth's atmosphere). Just as one expected (and indeed did see) a startlingly new and strange view of the universe when looking through the newly opened radio window, one should expect a radically new and strange view when "looking" through this newest window onto the universe—the neutrino window. Already we have seen a preview of coming attractions. For one thing, because of the unique ability of neutrinos to penetrate as can no other particle in the universe, we can "see" directly into the interiors of stars. Electromagnetic radiation, on the other hand, lets us see only skin deep, the outermost surface of most stars. The term "x—ray vision" is not applicable in astronomy because x—ray radiation goes no deeper down into a star than does ordinary light.

Neutrinos from Supernova 1987A were registered independently and unambiguously by two "water—Cerenkov" detectors specially designed to detect neutrinos. Ironically, each was designed and was operating at the time for a completely unrelated task. One, in a salt mine underneath Lake Erie, was searching for the theoretically—predicted proton decay. The other, in a mine beneath a mountain in Japan, was studying neutrino emission from the Sun. Both picked up neutrinos which had come from the first seconds following the almost—instantaneous collapse which spawned the explosion, but the teams of physicists operating both devices were totally unaware that the cataclysm had occurred. No one on Earth, astronomers or physicists or anyone else, knew what had happened until Ian Shelton took his fateful photograph of the night sky over Chile more than 24 hours later. The neutrino physicists, then, after the fact, examined their registers (which, fortunately, made permanent records) and found the trace of those neutrinos. Here we have an example of a serendipitous discovery: discovery of something unexpected made while in the process of looking for something else.

The very first neutrino physicists to realize that their detector might have picked up the neutrino flux, which astronomers had for years been predicting would accompany a supernova outburst, were those of the Swiss—French—Italian collaboration operating the installation beneath Mt. Blanc in the Alps. They were the first to announce detection, in I.A.U. Astronomical Circular No. 4323. They reported 5 "events" within 7 seconds at 2:52 A.M. (Universal Time) on February 23. Something was wrong, however, because that was almost 5 hours before the supernova exploded. The other two devices recorded their neutrinos simultaneously, within a few seconds, and no doubt did mark the true instant of the outburst.

Let us recall what actually happened, in the bowels of Sanduleak − 69° 202 once the collapse began, to let loose the flood of neutrinos. There were two distinct steps, although they were over and done with in about fifteen seconds.   (1) First, as a direct consequence of the unbelievably high density, protons and electrons (from disintegrated atoms in the star's core) were squeezed together to form neutrons, and, as a by—product, a type of neutrino called an electron—neutrino. Within a space of about 10 milliseconds, the blink of an eye, the neutron star was born of these new neutrons and approximately 10^{57} electron—neutrinos were produced as well. Immediately, the neutrinos made an attempt to escape *en mass* outward into interstellar and intergalactic space. If they made it, that would have been "the initial burst." What probably happened, however, was that the unbelievably high density of material in the new neutron star thwarted the escape, even though neutrinos are champions of penetration. (2) Next, as a consequence of the extremely high temperatures, probably up to 200 thousand—million degrees in the center, radiant energy was converted into mass. This, you might realize, is the exact reverse of the mass—into—energy reaction ($E = mc^2$) which powers atomic bombs and hydrogen bombs. Specifically, photons of the highest energy (gamma rays) are transformed into electron—positron pairs, which then almost immediately degrade into neutrino—antineutrino pairs. What, by the way, is an antineutrino? In the universe there exists or can exist an antiparticle that corresponds to every particle. If the regular particle, such as an electron, has an electric charge, then its antiparticle, in this case called a positron, will have the opposite charge. If the regular particle has no charge, as is the case with a neutrino, then the differences between it and its antiparticle are more subtle. Over and above the matter—antimatter distinction, there are three different types (particle physicists call them "flavors") of neutrinos: electron neutrinos, mu neutrinos, and tau neutrinos. Thus, in this so—called "thermal" phase, six different types of neutrinos are produced, the total number being 10^{58}, ten times more than were produced in the initial burst. These also have trouble plowing through the enormously dense neutron— star material, but they do gradually work their way out, in a process called "diffusion," over the next 10 seconds or so.

Once free of the dense neutron star material, the horde flew through interstellar (and eventually intergalactic) space at the fastest speed permitted in the universe, the speed of light. The extraordinarily thin interstellar and even thinner intergalactic medium (far, far thinner than the most nearly perfect vacuum on Earth) offered no resistance. Then, 170 thousand years later, those which were headed in exactly the right direction (only 10^{24} of the original 10^{58}) made their encounter with the Earth. Of those, all but a paltry two dozen either passed harmlessly through the water in the tanks of the Cerenkov detectors or else missed the tanks of water altogether and, for the most part, passed unimpeded through the Earth and continued their journey through interstellar space. It is those two dozen neutrinos we discuss next.

When a neutrino happens to interact with a particle of matter in a body of water, it does so by one of two mechanisms. (A) A neutrino of either of the three flavors can crash into one of the ten electrons which make up each water molecule, knock that electron off the molecule, and propel it more or less in the forward direction much as a cue ball sends one of the colored balls farther down the pool table. A particle moving at extremely high speed in a medium such as water will emit a characteristic blue light known as Cerenkov radiation, named after its discoverer, the Russian physicist Pavel Alexseevich Cerenkov (sometimes spelled Cherenkov). The electron moves only a few centimeters through the water before it slows down and stops radiating, but the flash of blue light it emits during its brief fraction—of—a—second flight is registered by one of many light—sensitive photoelectric cells surrounding and/or suspended within the tank of water. Each such flash of light duly recorded represents the detection of one neutrino. (B) An antineutrino of either of the three flavors can crash into one of the eight protons which make up each water molecule, be absorbed by it, and produce in its place a neutron and a positron. The positron then is propelled through the water at high speed, much as the electron was in the first mechanism, except that its direction can be in any random orientation, not preferentially in the forward direction. The same brief flash of blue Cerenkov radiation is emitted, however, and picked up by one of the photoelectric cells and recorded as a neutrino detection.

It turns out that detections by mechanism (A) are much less common than those by mechanism (B), even if the same number of neutrinos and antineutrinos actually sweep through the same tank of water. Moreover, electron neutrinos are more likely to interact than those of the other two flavors, the mu neutrinos and the tau neutrinos. Thus it will turn out that all but possibly two of the two dozen detections resulted from electron antineutrinos interacting *via* mechanism (B). This is somewhat unfortunate for two reasons. First, detections by mechanism (A), because they can yield information as to the direction from which the neutrinos came, give us effectively a neutrino telescope, not just a blind neutrino detector. Second, because only electron neutrinos are emitted during the important initial burst phase, all detections of antineutrinos must have come from the later diffusion stage, which is not capable of conveying as much information about a number of important questions which can be asked.

The I.M.B. detector is a roughly cube—shaped tank of purified water approximately 23 meters on a side lying 600 meters below the surface of Lake Erie about 40 kilometers east of Cleveland, near the town of Fairport, Ohio, in a salt mine of the Morton Salt Division of the Morton—Thiokol Corporation. The ten million liters of water are surrounded by 2048 photoelectric cells which look for the tell—tale flashes of blue light. The initials stand for the three institutions which collaborated to build the detector: the University of California at Irvine, the University of Michigan at Ann Arbor, and the Brookhaven (New York) National Laboratory. There were actually 36 individual scientists from 16 different institutions participating in the research project itself. Ironically the I.M.B. detector was designed (and was operating at the time) to search for something exceedingly rare but exceedingly crucial to our understanding of the fundamental nature of the universe: the decay of the proton. It had been calculated (according to one theory, at least) that, in a collection of 10^{31} protons, one single proton per year should spontaneously split apart into two fragments, a positron and another subatomic particle called a pion, both of which would speed off in opposite directions and produce the characteristic Cerenkov radiation which the detector would sense. The I.M.B. detector went into operation in 1982 and had yet to detect a single proton decay but, at precisely 7:35:41.37 (Universal Time) on February 23, 1987, it registered its first neutrino from Supernova 1987A, followed by 7 others within the next 6 seconds. As we explained above, all of the 8 were probably electron antineutrinos which diffused out of the hot neutron star during the first few seconds following its birth. Because one bank of photoelectric cells was out of commission at the time, the capability of detecting direction was severely compromised. Therefore it was not possible to decide experimentally which of the 8 events if any had resulted from detection of electron neutrinos by mechanism (A).

The Kamiokande II detector is a cylindrical tank containing six million liters of pure water sitting 900 meters beneath a mountain not far from the west coast of Japan inside a lead—zinc mine of the Kamioka Mining and Smelting Company. The six million liters of water are surrounded by 1071 photoelectric cells which also look for flashes of Cerenkov radiation. The detector's name is a combination of a place—name and an acronym: the Kamioka Nucleon Decay Experiment, and the Roman numeral II refers to a recent modification in the electronics to enable it to search more effectively for the neutrinos of relatively low energy which emanate from the Sun. Some 23 collaborators from six different institutions work together with this instrument. In its modified form, Kamiokande II began operation on January 1, 1986, and fortunately was in operation (looking for solar neutrinos) at the moment the first neutrinos from Supernova 1987A struck Japan. The first was registered at 7:35:35 (Universal Time) on the morning of February 23, 1987 and was followed by 10 others within the next 13 seconds. We cannot say with confidence that the first Japanese neutrino arrived six seconds before the first one which the I.M.B. detector picked up, because the time base at Kamiokande II was not calibrated precisely and could have been earlier or later possibly by as much as one minute. The relative spacing in time of their 11 neutrinos, however, was precise to a fraction of a second. Unlike the I.M.B. detector, Kamiokande II was able to determine, at least roughly, directions for the particles which produced the flashes of blue light. Thus, it was concluded, possibly two of their 11 neutrinos might have been electron neutrinos (not antineutrinos) detected by mechanism (A). Because those two were the very first two recorded by Kamiokande II, there is still considerable debate as to whether those two might have been electron neutrinos from the initial burst.

The original neutrino detector built to detect and measure neutrino emission from the Sun was also in operation when the 10^{24} supernova neutrinos swept past Earth. This detector, however, is not a water—Cerenkov detector, but is of quite different design and function. It is a cylindrical tank located almost one mile underground, in the Homestake gold mine near Lead, South Dakota. It contains 400,000 liters of perchloroethylene, a common dry—cleaning fluid, the chemical formula of which is C_2Cl_4. Approximately one of the four chlorine atoms on each molecule is of the proper isotope (^{37}Cl) to be able to interact with a neutrino, on rare occasions, and be converted to an atom of Argon (^{37}Ar). Interactions with solar neutrinos are so infrequent, despite the fact that 10 thousand—million pass through each square centimeter of the Earth's surface (and the tank of perchloroethylene) every second, that only a few interactions take place every month. At roughly monthly intervals, the scientists purge the tank of all gas which it might contain, including any argon atoms, which are in the form of a gas. Then, with the help of various techniques derived from nuclear physics, they actually count the number of individual argon atoms recovered. It is another story entirely to mention that, for the almost twenty years during which the experiment has been performed, the number of neutrinos measured has consistently been only about a third the number predicted by astronomers who think they understand the Sun's interior and its on—going nuclear reactions with confidence. But, getting back to Supernova 1987A, the Homestake detector was purged of its argon content after the month which included February 23, 1987 and no evidence of a single low—energy electron neutrino interaction (the only type it was capable of detecting) was found. That negative detection helped put an upper limit on the number of high—energy electron neutrinos which could have been emitted and confirmed the suspicion that neither the I.M.B. nor the Kamiokande II detectors had recorded any electron neutrinos from the initial burst. This is a good illustration of another thing that happens often in the day—to—

day execution of scientific research: a negative or null result, if carefully controlled and quantified, can nevertheless prove to be very useful.

Two other neutrino detectors in operation at the time, the one under Mt. Blanc mentioned earlier and the Baksam detector in the Soviet Union, searched their records for possible events related to the supernova explosion. The former found two and the latter found three. In both cases, the smaller number of counts is consistent with the fact that both of those detectors contain lesser volumes of water with which to capture neutrinos.

So, we have tagged $11 + 8 + 3 + 2 = 24$ neutrinos (probably all except two of them electron antineutrinos from the diffusion stage) from Supernova 1987A. Calculations based on (1) the volume of water in each detector, (2) the relevant detection efficiencies, and (3) the known distance between Earth and the Large Magellanic Cloud show that a total of 10^{58} neutrinos of all six types were emitted during the 10 seconds or so following the near–instantaneous collapse. Our Sun, in that same 10 seconds, gives off only 10^{38}. That may seem like a lot, but the supernova gave off 10^{20} times as many, during its brief moment of glory. In terms of energy, it can be calculated that Supernova 1987A broadcast into the universe 3×10^{53} ergs of energy in the form of neutrinos. One startling significance of that number is that it is 100 times greater than all of the energy used by the supernova to shine brighter than a thousand–million suns for several months, flood the universe with even more radiation in a form (like x–rays and radio waves) not as readily seen, lift off layers weighing more than the Sun and impart them with the escape velocity necessary to expand throughout and eventually leave the galaxy, and set a star (the new neutron star) spinning with rotational velocities on its surface equal to almost half the speed of light. That number 3×10^{53} ergs is even more startling in another respect, because it was given off in just 10 seconds. Brightness or luminosity is measured in terms of energy per second. In the visible or photographic portions of the electromagnetic spectrum, a supernova can become almost as bright as the entire galaxy in which it resides but rarely outshines it. If anyone had been looking with neutrino eyes on February 23, 1987, however, one would have seen a single star shine with the brightness of 10 million galaxies, for those glorious 10 seconds.

Almost as soon as the neutrino detection was made, scientists realized they could lay to rest one particular idea that had been floating around for a number of years. In an attempt to explain why consistently only 1/3 the expected number of solar neutrinos are detected, it had been suggested that 2/3 of them decay or die or disappear somewhere between the Sun and the Earth. Now we see that almost exactly the predicted number of stellar neutrinos from Supernova 1987A were detected. If virtually no neutrinos disappeared during the long journey from the Large Magellanic Cloud 170,000 light years away, how could 2/3 disappear during the short hop from the Sun only 8 light minutes away?

Shortly thereafter, scientists realized they had a golden opportunity to make another important measurement: the mass (if any) of the neutrino. Ever since the neutrino was postulated in 1930, it was supposed that probably it was a massless particle which travelled at exactly the speed of light. In the 50 years since then, however, there was the nagging doubt that perhaps the neutrino had a small mass and, consequently, did not travel with quite the speed of light. Various measurements in Earth—based laboratories had served only to put an upper limit on its mass, meaning it could weigh nothing at all or it could weigh anything up to that amount.

That the neutrino has mass has been an important possibility especially to cosmologists—astronomers who study the large—scale structure and eventual fate of the entire universe. Since the late 1920's it has been known that all of the galaxies are speeding away from each other, indicating that the entire universe is in a state of overall expansion. All of the mass in the universe and its mutual gravitational attraction acts collectively to try and retard this expansion. A central question in cosmology therefore has been: does gravity win and eventually stop the expansion or does gravity lose and leave the universe free to expand forever? More recently cosmologists have been paying attention to the new theory of "the inflationary universe" one aspect of which is that the contest between gravity and expansion should be a tie, with the expansion gradually slowing down and coming to a stop only infinitely far in the future, *i.e.*, at the end of time. However it turns out—win, lose, or draw—the important question is not yet settled to the satisfaction of the scientific community. One approach to trying to answer the question is to look at the average density of matter in the universe, which determines the strength of the gravitational retardation. The best estimate so far is 5×10^{-31} grams per cubic centimeter. The critical density corresponding to the borderline case is 10^{-29} grams per cubic centimeter. Thus, unless there is 20 times more mass than what we readily see, the expansion of the universe apparently will continue forever. Now the tiny neutrino enters the picture. Because neutrinos are probably the most plentiful particle in the universe, the so—called "missing mass" could be in the form of neutrinos if each one had only a small mass.

Recall that the 24 neutrinos from Supernova 1987A arrived within the space of 13 seconds, about what would have been expected from neutrinos originating during the diffusion stage and travelling the 170,000 light year distance at exactly the same speed. If they had any mass, however, their speeds would have been slightly different (depending on their energies, which were measured as a by—product of their detection) and their arrival times would have been spread out over longer than the 13 seconds which was observed. Looking at this quantitatively as best they could, scientists have estimated that neutrinos could not have had masses greater than about 10 electron volts. An electron volt is a unit of energy which particle physicists like to use to measure mass. For comparison, a single electron weighs 511,000 electron volts and a single proton (the nucleus of a hydrogen atom) weighs 938 million electron volts. If each neutrino really did have a mass of 10 electron volts, then the average density of mass in the universe would be equal to the critical value of 10^{-29} grams per cubic centimeter. But, because that 10 electron volt value is just an upper limit and the neutrino still might not weigh anything at all, the big question is still left hanging without a definitive answer. If any of the detections had picked up neutrinos from the extremely brief initial burst, then the sharpness of their arrival time compared to the expected arrival time (only about 10 milliseconds) could have been used to put a much tighter limit on the neutrino mass. Unfortunately, the initial burst from Supernova 1987A was not detected. We must wait for the next nearby supernova and plan to have perhaps even better neutrino detectors in operation at the time.

The most ambitious project afoot in the neutrino detection department is DUMAND, an acronym which stands for Deep Underwater Muon and Neutrino Detector. The idea was conceived in the 1970's and, in its most recent form, involves 756 photoelectric cells distributed over a $250 \times 250 \times 500$ meter volume of water (thirty thousand—million liters) 4.7 kilometers beneath the surface of the Pacific Ocean 40 kilometers west of Keohole Point on the Island of Hawaii. As with the other neutrino detectors already discussed, this plans to spot the flashes of Cerenkov radiation which result whenever a neutrino happens to interact with a proton or an electron in a water molecule. The information is to be carried underwater by fiber optics to an electronics laboratory on shore. Although it can respond to neutrinos of medium energy (around 10 million electron volts) such as the ones detected from Supernova 1987A, it is expected to be more useful when detecting the relatively rare neutrinos of extremely high energy (around one million—million electron volts) which might be coming from exotic objects somewhere in our galaxy or beyond. If very energetic neutrinos are detected with such a large—scale detector, scientists can determine the direction from which they came, with a precision of about one degree in the sky. Thus, DUMAND promises to be a very sensitive, multi—purpose neutrino detector and a bona fide neutrino telescope as well. This will be the first of the next—generation neutrino telescopes which we all hope will be in place and ready to respond when the next supernova explosion detonates in our own galaxy.

CHAPTER 15

THEORISTS

INTRODUCTION

In astronomy, as in many other sciences, there is somewhat of a distinction between theorists and observers. While some scientists successfully straddle the fence between these two camps, most consider themselves as tending towards one group or the other.

Often observers come up with observations that will not fit within the confines of current theories. This can suggest the need for a new, improved theory. It is always the delight of an observationalist to "show the way" in this regard.

It is the theorist's job to develop and refine theories—often mathematical models—based on the available observations. A theory is particularly useful when it explains all the current observations, and also predicts what would be found if some new type of observation were to be made. When such predictions are made, observers often scramble to be first to make the observations the new theory predicts. If the observations are as predicted, the theorists, with a smug grin, says "Of course that's what you observed, didn't I tell you that's what you'd find?" If the new theory is not confirmed with the new observations, the observationalists tell the theorists, "Another wild theory blown out of the water by a good, real world, empirical observationalist!"

In the case of Supernova 1987A, the interplay and competition between the theorists and observationalists has been real and intense. At one meeting (see Chapter 16), the more vocal members of the two groups stood up and shouted at each other! "We predicted that was what you would observe, why didn't you believe us?" and "Now here is an observation you theorists will never be able to explain!"

Theorists, who must show that their mathematical theories fit the observed facts, get involved in serious number crunching. Various parameters (coefficients) in the models are "adjusted" to make the model fit the observations. (No, this is not considered cheating if done properly.) In the early days of science, theorists had to do their calculations by hand. This tended to keep the models simple! At a later stage, rooms full of Frieden electro—mechanical calculators and "computers" (the people that operated the calculators) allowed the theorists to invent fancier models and still do all the calculations needed to see if they could be adjusted to match the real—world observations.

However, it was the coming of the large digital computers that really changed how theorists worked. Successively more powerful computers removed the barriers to model complexity, and the theorists were able to run almost rampant. In times past, a "model" was usually a small set of mathematical equations, typically written down in a few lines, that described some physical phenomena. These equations could be published, examined, and thought about by many. The situation is very different now. The heart of things for theorists these days is a program containing many lines of instructions for the computer. Top—of—the—line supernova programs run to 30,000 lines. Such a complex program has to be developed in stages and debugged and refined over time. The program "evolves" over time, increasing in sophistication.

Faster and more powerful computers allow the theorists to either run bigger, fancier models, or to make more frequent runs on the same sized models. It is not surprising that top theorists for the past couple of decades have, in the main, been associated with institutions that can afford to own and operate the biggest computers. This is, perhaps, finally beginning to change. Microcomputers, which are affordable by even the smallest institutions, are becoming so powerful that some theorists are beginning to use them.

Astrophysics in general, particularly the astrophysics of supernovae, is closely tied to nuclear theory—the interaction of atomic particles. Nucleosynthesis, the formation of atomic nuclei as a result of nuclear reactions, is particularly central to supernovae. George Gamow, when he suggested his "big bang" theory, thought that all the elements could be created in the big bang itself—the heavier elements being built up from the lighter ones. Willy Fowler and others were able to show, however, that this was not possible, that there were some "gaps," elements that could not be made or bridged across. Fred Hoyle suggested that the heavier elements might be synthesized in the interiors of stars, as opposed to the big bang. Geoffrey Burbidge, Margaret Burbidge, Fowler, and Hoyle crystallized these ideas in the late 1950's. It became clear over time that supernovae played a particularly crucial role in the synthesis of the elements, particularly those heavier than iron.

It is not possible in a short chapter, or probably even in an entire book, to do justice to the many theorists currently involved in supernova research. Even if it were possible, it would not be appropriate to do so, for we only wish to give the "flavor" of this aspect of the Supernova 1987A story. Thus, we discuss only a few theorists, and purposely a heterogeneous group at that.

The first two astrophysicists, we describe, Willy Fowler and Stan Woosley, are clearly "related." Stan is Willy's "grand student." Jim Truran is one generation removed from Al Cameron, who independently of Willy Fowler worked out the basic ideas of nucleosynthesis in stars. Ken'ichi Nomoto is the "grand student" of C. Hayashi, the Japanese expert on stellar evolution. Roger Chevalier has successfully modeled a number of aspects of supernovae and their remnants, and has, on occasion, worked closely with observationalists, such as Bob Kirshner (whom we met in Chapter 12). Virginia Trimble is not a pure theorist, as her work on supernovae has covered a number of areas. Stirling Colgate did important, early computer modeling of supernovae, pioneering work of which any self—respecting theorist would be proud. However, as we saw in Chapter 5, Stirling, unlike other theorists and many observers, thought computers should be applied to the problem of observing— supernovae in this case. Finally a nova theorist, Sumner Starrfield, is profiled.   There are important connections between novae and supernovae—especially Type I supernovae.

WILLIAM A. FOWLER

William A. Fowler was born in 1911 in Pittsburgh, Pennsylvania, but moved to Lima, Ohio, at the age of two when his father, an accountant, was transferred to Lima from Pittsburgh.   Lima was a railroad center served by the Pennsylvania, Erie, Nickel Plate, and Baltimore & Ohio railroads.   It was also the home of the Lima Locomotive Works, which built steam locomotives. As a boy, Willy, as he was nicknamed later, spent many hours in the switch yards of the Pennsylvania Railroad not far from his home. Willy attended Lima Central High School, and was President of the Senior class of 1929.

Upon graduation he enrolled at The Ohio State University in Columbus, Ohio, in ceramic engineering. Fortunately all engineering students took the same basic courses, including physics and mathematics, and Willy became fascinated with physics. A new degree was offered in Engineering Physics, and Willy enrolled in that option at the start of his sophomore year.

Upon graduation from Ohio State, Willy went to Caltech as a graduate student under Charles Christian Lauritsen in the W. K. Kellogg Radiation Laboratory. Charlie Lauritsen supervised Willy's doctoral thesis on "Radioactive Elements of Low Atomic Number." After he received his doctorate, Willy stayed on at the Kellogg Radiation Laboratory. Charlie's son, Tommy Lauritsen, did his doctoral work under Charlie and Willy, and the three of them worked together as a team for over thirty—five years. They were primarily experimentalists. In the early days, Robert Oppenheimer taught them the theoretical implications of their results.

Figure 1.   William A. Fowler, Professor of Physics (Emeritus) at the Kellogg Radiation Laboratory of the California Institute of Technology. Willy's work at Caltech was instrumental in establishing that stars could synthesize lighter elements into heavier elements.   Willy considers Stan Woosley his "grand student" via Don Clayton.

Hans Bethe's announcement of the CN–cycle in 1939 changed their lives. They were also studying the nuclear reactions of protons with the isotopes of carbon and nitrogen in the laboratory, the very reactions in the CN–cycle.   Although World War II intervened, the Lauritsens and

Fowler were able to quickly restore Kellogg as a nuclear laboratory after the war, and decided to concentrate on those nuclear reactions which take place in stars. They called it Nuclear Astrophysics. Before the war Hans Staub and William Stephens had confirmed that there was no stable nucleus at mass 5. After the war, Alvin Tollestrup, Charlie Lauritsen and Willy confirmed that there was no stable nucleus at mass 8. These mass gaps spelled the doom of George Gamow's idea that all nuclei heavier than helium (mass 4) could be built by neutron addition one mass unit at a time in his big bang. Edwin Salpeter of Cornell came to Kellogg in the summer of 1951 and showed that the fusion of three helium nuclei of mass four into the carbon nucleus of mass twelve could probably occur in red giant stars but not in the big bang. In 1953, Fred Hoyle induced Ward Whaling at Kellogg to perform an experiment which quantitatively confirmed the fusion process under the temperature and density conditions which Hoyle, Martin Schwarzschild, and Allan Sandage had shown occur in red giants.

The grand concept of nucleosynthesis in stars originated with Hoyle as early as 1946. After Whaling's confirmation of Hoyle's ideas, Willy became a believer and in 1954/1955 spent a sabbatical year in Cambridge, England, as a Fulbright Scholar in order to work with Hoyle. They were joined by Goeffrey and Margaret Burbidge. In 1956 the Burbidges and Hoyle came to Kellogg and in 1957 their joint efforts with Willy culminated in the publication of "Synthesis of the Elements in Stars" in which they showed that all of the elements from carbon to uranium could be produced by nuclear processes in stars starting with the hydrogen and helium produced in the big bang. This classic paper has come to be known from the last initials of the authors as B^2FH. A. G. W. (Al) Cameron single—handedly developed the same broad ideas at about the same time.

Experimental measurements of the cross section (the probability of occurrence) of hundreds of nuclear reactions and their conversion into stellar reaction rates had been necessary so that nucleosynthesis in stars could be quantitatively confirmed. The Kellogg Laboratory and Willy Fowler played a leading role for many years in this effort, and for this he shared the Nobel Prize for Physics in 1983.

STANFORD E. WOOSLEY

One of the premier computer programs for modeling the evolution of massive stars and their explosions as supernovae is named KEPLER. It was named not after the astronomer but after the last bright naked eye supernova until now (which was discovered by Kepler). Developed (initially coded) by Stan Woosley and Tom Weaver some 14 years ago, the program now has over 30,000 lines of instructions. The code has been continuously refined over the past 14 years by Stan and Tom (the Simon and Garfunkle of supernovae modelers). A run of the model takes between 30 and 100 hours of computer time on a control Data Corporation 7600 computer—one of the largest and fastest computers in the world. At "prime time" rates, 100 hours of CDC 7600 time would cost a small fortune, and many different "runs" are needed. To get around this, the runs are made in a low cost background mode. When the CDC 7600 is not busy doing some priority job, it works on one of Stan and Tom's models. There are two CDC 7600 computers at the Lawrence Livermore Laboratory which Stan and Tom use. They usually have a model running on each computer, and it is often several weeks before runs are completed. Stan now does many of his calculations on two Microvax computers in Santa Cruz.

It should be noted that Stan, always full of good humor, has a dog named Kepler, a nine year old Sheltie, also named after the computer program. Stan points out that when Kepler circles the dining room table, which he does frequently, he keeps the square of his period proportional to the cube of his distance from the food.

Stanford E. Woosley was born in Texarkana in 1944. Stan had an early interest in science, particularly chemistry and physics, and space had a special appeal. Young Stan thought traveling to Mars or the stars would be keen. While many sharp young boys had similar interests, Stan's interest in explosions set him apart.

Stan not only studied explosions, but he built bombs. Fortunately, Stan's bombs never became very powerful and he was not a public nuisance. They were never more sophisticated than could be made from drug store supplies. (In the good old days, the drug stores had potassium nitrate, potassium permanganate, sulfur and the like, sometimes one could even get a little potassium chlorate.) Chemistry was fascinating to Stan. The idea of things reacting with other things to make new things was intriguing. The fact that hydrogen (which one could get by mixing kitchen lye with aluminum foil) would burn in air to make water, which one usually used to put out fire, was amazing. Stan survived all his bombs, and graduated from high school in Fort Worth, Texas, and entered Rice University in Houston, Texas, in 1962. Later when Stan began to work on the origin of the elements by nuclear reactions in stars, a subject to which he was introduced by Don Clayton at Rice University, he found, with the same fascination, that carbon could come from helium and iron from carbon. That is really why Stan became interested in supernovae, so he could study the origin of the elements in these big explosions.

Stan ended up staying at Rice for 10 years. He received his BS in Physics in 1966, and his PhD in Astrophysics in 1971. At Rice University, Stan was the cross—product of two famous "nucleosynthesis families." One was the William Fowler/Donald Clayton line, while the other was the A. G. W. Cameron/W. David Arnett line. Clayton was Stan's thesis advisor, while Arnett was his assistant thesis advisor. Stan's thesis was on "The Explosive Burning of Oxygen and Silicon." Stan's bombs were really big now—star—sized.

Figure 2.   Stanford E. Woosley, Professor and Chairman of the Department of Astronomy and Astrophysics at the University of California at Santa Cruz.   This picture was taken during a press conference at the Lawrence Livermore National Laboratory following a meeting of the "Northern California Supernova Dealers Association."

After receiving his PhD, Stan stayed on at Rice as a post—doc for a year and a half. It was during this time that he spent the first of three summers with Fred Hoyle and other astrophysicists in Cambridge, England. Fred's Institute for Theoretical Astrophysics was at the cutting edge of theoretical astrophysics, and many of the best in the field came for these summer gatherings. The nature of the Institute changed when Fred resigned his professorship. The Institute still exists and is now called the Institute of Astronomy. Martin Rees has been its director for many years. It still is a major gathering place for astronomers and astrophysicists. For example, Stephen Hawking of black hole and Hawking radiation fame is there. However the participation of Clayton, Fowler, and Woosley was greatly diminished after Fred resigned.

Figure 3.   W. David Arnett in the fall of 1986, with Hans Bethe in the foreground.

For the second part of his post–doc, Stan was accepted as a Research Fellow in Physics at the Kellogg Radiation Laboratory of the California Institute of Technology, in Pasadena, California. This was the bastion of Willie Fowler, and Stan spent the next two and a half years doing pure nuclear reaction rates. As implied by their name, nuclear reaction rates are how long it takes a reaction between atomic particles and/or atoms to occur (on the average) under some set of physical conditions (temperature, pressure, *etc.*). Accurate reaction rates are vital to models of nuclear reactions as in atomic bombs, normal stars, and, of course, supernovae.

With his Physics Fellowship at an end, Stan accepted a position with the University of California at Santa Cruz. Located in a stand of redwood trees on the California coast, UC Santa Cruz is one of the most beautiful campuses in the world. It is easy to understand why Stan still works at Santa Cruz, where he has risen to full professor and Chairman of the Board of Studies in Astronomy and Astrophysics.   Stan, unassuming, informal, and still boyish looking, is not what one thinks of when the terms "full professor" and "Chairman" are mentioned. Stan is quick to smile, like Bob Kirshner, and has much the same boyish energy, impishness, and light touch.   To them, research is pure fun, and they show it.

Figure 4.   Thomas A. Weaver, Stan's partner in the development of Kepler, the computer code that models supernovae explosions.   Tom works in the Physics Department, Special Studies Group, at the Lawrence Livermore Laboratory.   The picture was taken at Cornell University in the Fall of 1986.

Besides being on the staff at Santa Cruz, Stan became a consultant at the Lawrence Livermore Lab (LLL). Tom Weaver was at LLL, and Stan and Tom soon set out to put all the then known physics about stars into a computer model to run on the mainframe computers at LLL. While the lab was well known for its work on weapons, it also supported pure research, such as astrophysics.   Of course, nuclear reactions in bombs and stars have many similarities, and in both cases they are studied *via* complex computer models.

Figure 5.   Left to right: George Michael Fuller, Enrico Fermi Institute, University of Chicago, Stanford E. Woosley, University of California at Santa Cruz, and James R. Wilson.   George provided the "equations of state" for the initial model.   Jim runs the core collapse and rebound (or black hole).   The picture was taken at Cornell University in the fall of 1986.

Stirling Colgate's early computer modeling of supernovae was done on the computers at Livermore, and later at Los Alamos National Laboratory—another research center where nuclear reactions for both bombs and stars are studied.   It is also interesting to note that Robert Oppenheimer was at Berkeley with Lawrence Livermore, and then founded the Los Alamos Lab when he was in charge of developing the first atomic bomb.   Oppenheimer made important early theoretical predictions regarding supernovae and neutron stars.

Stan and Tom enlisted George Fuller to provide the "equations of state" that define a basic star and the changes that happen in stars.   The extremely sudden collapse of the iron core, formation of a neutron star, and the rebound (or the formation of a black hole in some cases) was too much for their model, and a collaboration with hydrodynamicist, Jim Wilson, began. Stan and Tom would run their model to the point of core collapse, and then Jim would run his to the neutron star (or black hole) point, and then Stan and Tom would finish the run.

Figure 6.   Left to right, Stan Woosley, Lisa Ensman, and Phil Pinto in Stan's Office at the University of California at Santa Cruz.   Phil joined the team in 1984 and has primarily worked on the spectra of Type I supernovae.   Lisa joined the team in 1985, and has worked on the light curves of massive stars that have lost much of their hydrogen envelopes.

Additions were made to the model to model a supernova spectrum, and later, to model a light curve. They were not the first theoretical models of supernova light curves, as Falk and Arnett had modeled them earlier, as had Nadĕzhin and a number of co-workers in the Soviet Union, as had Colgate. The light curve work was done by Weaver and

Stan.   Tim Axelrod entered as they began to study Type I as well as Type II supernovae.   Axelrod did his thesis on the spectra of Type I supernovae and was the first to point out the spectroscopic evidence for radioactive cobalt in Type I supernovae.   Phil Pinto has also worked primarily on the spectra of Type I supernovae.   Only very recently, since 1987A, have he and Axelrod begun to work on the spectra of Type II supernovae.   Lisa Ensman is the latest addition.   She works on the light curves of massive stars that have lost much of their hydrogen envelope. Pinto is presently a post–doc at UCSC.   Ensman is a graduate student and Axelrod is now a full time employee at Livermore.   The light curve work became an important comparison between theory and observation when Supernova 1987A exploded because of the early discovery and complete light curve coverage of this bright supernova.

February is warm and pleasant in Santa Cruz, but at Lake Tahoe the snow is deep.   Stan was thinking about a 10–day skiing vacation at Lake Tahoe, and he was about to leave on when Phil Pinto handed him the IAU Circular sent out by Brian Marsden (see Chapter 1).

Stan's immediate reaction was that it was just too good to be true—they would take it away.   He remembered the "announcement" of a supernovae—a joke that got a bit out of hand.   But the phone started ringing.   Everyone was calling everyone else.   Phone bills grew asymptotically.   It really was true!   It was like Christmas.

Stan felt Brian Marsden was the real hero of the first week or so of 1987A.   The IAU circulars were carefully done and very informative. The astronomers anxiously awaited the arrival of the latest circular.   Lick Observatory, along with lots of other observatories, quickly signed up for the much faster computerized transmittal of the IAU circulars.

After the initial elation wore off, Stan realized he was supposed to be doing something!   Phil Pinto and Peter Martin were enlisted (Tom Weaver was on vacation), and the first 1987A model was launched.   The first paper was received at the *Astrophysical Journal* on March 3.   It was the first paper submitted on the subject of 1987A to any journal.   They predicted in this first paper that the light curve should brighten shortly owing to radioactive energy deposition.   They also pointed out that the timing of the supernova signal and first optical acquisition indicated a blue progenitor and not a red one.   Interestingly in accepting the paper, the referee asked that they change their paper to account for 1) the fact that SK −69° 202 had not really exploded, and 2) the predicted radioactive brightening had not occurred.   By the time they got their revised paper sent to the journal the brightening was in progress and shortly thereafter the IUE team identified SK −69° 202 as the progenitor. As new observations became available, the parameters of the 1987A model were refined, and new predictions were made.

At times, results from the observationalists that strongly conflicted with the model were questioned or called out as wrong.   Such was the case when NASA called a meeting about two weeks after 1987A went off. Bob Kirshner announced that it was not Sanduleak −69° 202 that went off.   Stan disagreed, of course.   He had already stated, and in writing in the paper sent to the *Astrophysical Journal*, that model said it had to be

Sanduleak −69° 202 or a star very much like it (and there was no similar one in the area).

A second paper was received from Stan by the *Astrophysical Journal* on April 22nd, with co−authors Phil Pinto and Lisa Eisman. A third paper was presented at the Minnesota conference "The First 100 Days of Supernova 1987A." Stan confidently predicted in this paper what the supernova would do in its first 300 days. So far he has been right on track.

And what does Stan think about all this? He feels that 1987A is going to be with astronomy for a long time.

"The neutrino burst was absolutely fantastic. The Japanese and IMB folks should get a Nobel Prize. As things clear up we'll be able to look right into where the new elements were forged in the oven, and see down to the neutron star—it is there!"

JAMES W. TRURAN

James W. Truran was born in 1940. He had a general interest in astronomy based on many discussions with his father. After graduating from Cornell University in 1961 with a major in Physics, Jim became actively interested in astrophysics during his first year of graduate school at Yale, following conversations with Al Cameron. Dave Arnett (now at the University of Chicago), Carl Hansen (now at the University of Colorado), and Jim all began working with Al on astrophysical problems at that time.

Throughout his career, Jim's research interests have remained concentrated in the general area of nuclear astrophysics, including studies of thermonuclear reaction rates, nucleosynthesis, stellar evolution, supernova explosions, nova explosions, and galactic chemical evolution. In recent years, an increasing fraction of his time has been spent specifically in studies of the nature of the outbursts of classical novae.

Jim's thesis title, "Thermonuclear Reactions in Supernova Shock Waves," makes clear his early interest in problems associated with nucleosynthesis accompanying supernova explosions, or "explosive nucleosynthesis". The physics of his thesis was concerned with theoretical calculations of nuclear reaction rates for conditions compatible with stellar and supernova environments. The main problem he pursued was "explosive silicon burning" and related problems. These studies subsequently led to the recognition that the decay of nickel provides a critical energy source powering the light curves of Type I supernovae and perhaps the later stages of the light curves of Type II supernovae as well. It might be noted that Jim's calculations describing explosive nucleosynthesis, as well as those of his graduate student colleagues at Yale (Arnett and Hansen), were inspired by Al Cameron's intuition that such explosive environments were the likely source of the bulk of the nuclei from neon to nickel.

Jim's interests in the nature of the progenitor of Supernova 1987A were motivated as well by calculations performed by one of his former graduate students, Wendee Brunish.   Wendee's calculations of the evolution of massive stars of about 15–30 solar masses of low metallicity indicated a significantly slower redward evolution as a supergiant than is typical of stars of normal (solar) chemical composition.   This may account for the presence of a blue supergiant progenitor (Sanduleak $-69°$ 202) for Supernova 1987A.

What does Jim think about 1987A?

"A nearby supernova is an event that most supernova theorists only dared to dream might occur in their active research life time.   I find it intellectually exciting: I find it a tremendous amount of fun.   I have thoroughly enjoyed the interactions regarding Supernova 1987A that I have had with my colleagues and collaborators here in Munich: for me, this has been a spectacular sabbatical year.   And we should remember that the fun and excitement is by no means over. There may be further surprises!   We still await the opportunity to probe, among other things, the nature of the condensed remnant (neutron star or black hole) and the composition of the innermost layers of matter ejected by this supernova.   Modeling of the light curve continues to provide challenges.   The fun can continue for years.   Our past knowledge of supernovae was gained either from observations of outbursts in distant galaxies (too far away to hope either to receive significant neutrino fluxes or to follow the spectral evolution past the first year or two), or from studies of old remnants in our galaxy.   Observations of Supernova 1987A can and certainly will continue for years.   Much is yet to be learned."

Jim has been a Professor of Astronomy at the University of Illinois since 1973.   He will return to Illinois when his sabbatical in Germany is completed.

KEN'ICHI NOMOTO

Ken'ichi (Ken) Nomoto is currently an Associate Professor of Astronomy of the University of Tokyo.   He received his doctorate at the University of Tokyo, based on calculations of the pre–supernova evolution of massive stars.

After Ken received his doctorate, he started work on studies of accreting white dwarfs in relation to novae and supernovae.   While staying at NASA's Goddard Space Flight Center for two years (1980–

1981), he had many opportunities to attend meetings and get to know other astronomers.

Figure 7.   Willie Fowler, Ken Nomoto, and Jim Truran at a meeting at Yerkes Observatory, Green Bay, Wisconsin, in 1983.

At a meeting in 1980 in La Jolla, Ken was told that the carbon deflageration model which he had developed in Japan might be a good model for Type I Supernovae.   This encouraged him to study these objects further.   Other important information came from the astronomers who obtained CNO abundances of the Crab Nebula utilizing the International Ultraviolet Explorer.

In 1982–1983, Ken had a chance to stay at the Max Planck Institute in Munich for several months with collaborators Wolfgang Hillebrandt and Friedel Thielemann.   He made an explosion model for 8–10 solar mass stars (with Wolfgang) and completed detailed nucleosynthesis calculations for Type I supernovae (with Friedel).   Ken's collaboration with Dave Branch on theoretical spectra was crucial to showing that Ken's carbon deflagration model of an accreting white dwarf could account for many features of Type I supernovae.

In 1986, Ken visited the State University of New York at Stony Brook and Brookhaven National Library where he worked with Gerry Brown and Sid Kahana on pre–supernova models of massive stars, *i.e.*, progenitors of Type II supernovae.   By improving the physics of the model, and the details of the nuclear reactions, he found that the massive stars form significantly smaller iron cores than in earlier models by Stan

Woosley and the Livermore group. This is crucial to the mechanism that transforms collapse into explosion, a mechanism very sensitive to the mass of the iron core. Such a small iron core as Ken obtained would lead to a very energetic explosion.

Figure 8.   Ken Nomoto and Willy Fowler in 1982 at the latter's office at the California Institute of Technology (Cal Tech) in Pasadena, California.

This is important in modeling Supernova 1987A, because the very sharp rise in the visual luminosity in the three hours after the neutrino burst requires a large expansion velocity and thus a large explosion energy. Such a large explosive energy could not be obtained from the so—called "delayed" neutrino heating mechanism found by Jim Wilson, currently a very popular model. Ken thinks that the unusually sharp rise in brightness points to the formation of the small iron core in the progenitor and the "prompt" explosion due to rebounding shock.

ROGER A. CHEVALIER

Roger first become interested in astronomy at the age of ten, when he used a small telescope to look at the moon and planets. He has pursued astronomy from then until the present.

In high school, Roger built an 0.2—meter reflecting telescope. At Caltech, a heavy emphasis on physics and mathematics encouraged Roger to go into theory.   At Princeton, his PhD thesis, under the direction of Jeremiah P. Ostriker, involved the evolution of supernova remnants.

Figure 9.   Ken Nomoto and Wolfgang Hillebrandt in May 1983, in Erice, Italy.

Roger went to Kitt Peak National Observatory in 1973.   Robert Kirshner was a postdoc at Kitt Peak from 1974 to 1976.   Bob had, as noted in an earlier chapter, recently finished his PhD thesis on spectrophotometry of supernovae, and the new data greatly clarified the observational aspects of supernovae.   Since Roger had a background in explosive phenomena, he decided to model the supernova light output.

Roger went from Kitt Peak to the University of Virginia in 1979, where he is currently Chairman of the Astronomy Department and Director of the Leander McCormick Observatory.   Some of Roger's key contributions include:

1.   The development of models for the hydrodynamics and light curves of Type II supernovae which showed that the basic properties of their events follow naturally from the explosion of a massive star at the end of its life. The light curve does not depend on exactly how the initial impulsive energy is deposited.

2.   He showed that a massive star that has undergone considerable mass loss ends its life with a relatively small radius and is fainter than a normal supernova.   This suggestion is particularly relevant to Supernova 1987A.

3.   Roger (along with Bob Kirshner) showed direct evidence for stellar nucleosynthesis in the fast knots ejected in the Cas A supernova. Different knots appear to have undergone different amounts of nuclear burning.

4.   He identified the carbon deflageration of 1.4 solar mass white dwarfs as the most likely source of Type I supernovae. This is currently the preferred model for the Type I events.

5.   He modeled the radio and X—ray emission from Type II supernova as the result of interaction with the circumstellar medium. This model explains the radio emission observed from Supernova 1987A.

VIRGINIA L. TRIMBLE

Virginia Trimble was born in 1943, in North Hollywood, California. Her father, a research chemist, still occupies the house where she was born. Virginia's early exposures to astronomy included a maternal uncle who was a serious amateur astronomer, though she generally regarded having to look through Uncle Roy's telescope as a tiresome experience (probably due to extreme uncorrected myopia), frequent visits to Griffith Observatory and Planetarium (but also to the zoo, local amusement park, hiking grounds, *etc.* reflecting her father's view that it was his duty to get Virginia out of her mother's hair one afternoon a week), and copious reading of elementary and popular astronomy books (but also of paleontology, linguistics, archaeology, and many other things).

Virginia entered UCLA shortly after her 17th birthday as a declared astronomy—mathematics major, but was assigned George Abell, well—known for his observations of galaxies and for his textbooks, as her undergraduate advisor. Virginia completed her graduation requirements at the end of academic year 1963—64. On advice the of Abell, she applied to his graduate institution (Cal Tech) and to no other, and for fellowships from Woodrow Wilson Foundation (awarded) and National Science Foundation (not awarded; first year women did not get them in those days).

Caltech had then a policy that women students could be admitted only under "exceptional circumstances" (at the graduate level; there were no undergraduate women until well into the 1970's). On the other hand, Virginia would be coming with her own scholarship money (the "exceptional circumstance" she claimed was that Woodrow Wilson Foundation required recipients to change institutions from their undergraduate one and there were no other astronomy programs within driving distance of home). The letter of admission from CIT said "Dear Miss Trimble: We have reviewed your applications materials and, on the basis of this, we cannot deny you admission to Caltech. We feel, however, on the basis of your interests, that you would be happier elsewhere."

In spite of this, Virginia went to Caltech and was supported for the second year on a research assistantship from Jesse Greenstein's research grant and on National Science Foundation Fellowships for years three and four. Thus Virginia worked with Guido Munch, who gave her the task of measuring positions, proper motions, and radial velocities in the Crab Nebula (on existing plates taken by himself and the late Walter Baade) and measuring line strength ratios from plates she was to take herself on the 48—inch Schmidt at Palomar Observatory. She became the second woman ever assigned observing time in her own name at Palomar (the first was Vera Rubin). The Crab Nebula in 1966 was beginning to sound exciting because it had recently been identified as the first X—ray source outside the solar system (and as the first radio source about 10 years before), and as having a compact, radio source at its center. The goals were to determine expansion rate, distance, energetics (acceleration; excitation; ionization, *etc*), three—dimensional structure, and, with luck, the center of expansion well enough to pick out the energy source. Although measuring proper motions on plates from short—focal—length telescopes was then widely regarded as nearly impossible, the project proved perfectly doable and resulted in a completed thesis in slightly less than two years.

Virginia submitted her thesis in April 1968, the discovery of pulsars having been announced in March. A close association with supernovae and the previously—hypothetical neutron stars was immediately assumed by almost everybody. In late November/early December, the presence of the fastest—to—date pulsar was confirmed in the Crab Nebula, which instantly made it a very exciting object to a wide range of astronomers. This was a true piece of serendipity, and resulted in Virginia being asked to speak many more places than a fresh PhD normally would, and this undoubtedly contributed to her always having job offers when she needed them.

Virginia's thesis, "Motions and Structure of the Filamentary Envelope of the Crab Nebula," showed that the expansion was quite radial and regular. Given the date of the Chinese record, she showed that the expansion had been accelerated (requiring a continuous input of energy comparable with that needed to maintain the X—rays, and now known to come from the slowing down of the pulsar). The center of the expansion was not at present at the geometrical center of the nebula and was not on top of any of the visible stars near the nebular center (a little later, this became one of the early pieces of evidence that pulsars are high velocity stars). The distance was found to be about 65% larger than earlier estimates.

Trimble has continued her work on supernovae, concentrating on supernova remnants. She has published many papers in the field, although it is just one of her many interests.

Virginia's Reactions to 1987A

"The obvious emotions of 'isn't it exciting' and so forth (the more so as one of the neutrino groups is headquartered here at University of California at Irvine, and my husband, Joe Weber, was operating two of the three gravitational radiation antennae that were collecting data at the time of the explosion). Undoubtedly we will learn all sort of neat things, and supernova research will benefit enormously in intellectual, political, financial, and other ways. Members of the Supernova Working Group of the International Astronomical Union (whose chairmanship I will shortly and gratefully hand over to William Liller) have expressed an assortment of regrets—that it wasn't in the northern hemisphere; that something—or—other (Space Telescope or whatever) wasn't ready; that it wasn't a Type I so that the distance scale could be calibrated; and so forth." I guess I slightly regret, from the point of view of supernova research, that it wasn't a more conventional Type II of the kind that dominates catalogues and (we think) contributes most to nucleosynthesis. But we should not be surprised—a couple of years ago, I. S. Shklovskii published a short note suggesting that the reason we had never seen a Type II supernova in an irregular galaxy (of which the LMC is a standard example) is that, with the lower metal abundance, massive stars never developed the extended, cool, red supergiant envelopes needed to make the standard Type II Supernovae light curve. I was always pleased that Fritz Zwicky lived to see the discovery of a neutron star in the Crab Nebula (which he had predicted in 1934) and to say "I told you so." I am very sorry indeed that Shklovskii did not live to say 'I told you so' on this one."

STIRLING A. COLGATE

Born in 1925 in New York City, Stirling A. Colgate grew up in both New Jersey and New York State—attending Scarborough and Far Hills Schools. Stirling had an early interest in mechanical and electrical machinery that later evolved into an interest in math and science. Stirling graduated from Los Alamos Ranch School at the end of 1942 when the Manhattan Project took it over for the nuclear weapons laboratory. Stirling recognized Oppenheimer and Lawrence and correctly concluded that the fission ratio (number of neutrons per fission) was indeed greater than two and therefore the new Los Alamos laboratory must be for nuclear research. He spent a year at Cornell before joining the Merchant Marine in WW II, both as electrician and second engineer. He returned to Cornell changing from engineering to arts and science in order to study physics. Stirling developed several experiments for the undergraduate laboratory. One was an alpha particle range experiment, which allowed him to keep close touch with the laboratory, his first love.

Upon receiving a BA in physics in 1948, Stirling decided to stay on at Cornell for graduate school. He worked for Robert Wilson and did a thesis on the experimental measurement of gamma ray absorption coefficients. Stirling had made his decision to go into physics while in the Merchant Marines; after hearing of the explosion of the nuclear bomb he understood the bomb well enough to inform the crew of his merchant ship at some length. Consequently, upon receiving his PhD he chose between Oak Ridge, Los Alamos, and the Radiation Laboratory at Berkeley to work with Louis Alvarez at the latter on the newly–developing giant accelerators for producing fission power. After a year he moved to Livermore with Herb York to start the early work in thermo–nuclear fusion controlled by magnetic fields (nothing to do with nuclear weapons). However, it soon became apparent that the Laboratory started by Edward Teller would develop and test nuclear weapons. Stirling was in charge of the nuclear diagnostics on large thermo–nuclear weapons tests in Eniwetok and Bikini. It was not until five years later, after spending a year in Geneva with the State Department, that discussions with the Soviets on detecting nuclear weapons in space spurred Stirling's imagination to consider what a supernova would look like. Consequently he became more of a theorist, although in those years he was working as a plasma physicist in controlled fusion experimental work. His early work on supernova theory with Dick White was strongly encouraged by Willy Fowler and Al Cameron.

Stirling started his theoretical work on supernovae in 1960 with Montgomery Johnson by trying to model the external energetic blowoff from a supernova explosion as a likely mechanism for the origin of cosmic rays. This was analogous to phenomena associated with nuclear weapons in space. It was later in 1962 with Dick White that they decided it was necessary to understand how a supernova really works if you are going to invoke it to make cosmic rays. This work was completed in 1964, but not published until two years later because of the difficulty of publishing at secret weapons laboratories.

In early 1965 Stirling went to New Mexico Institute of Mining and Technology (NMIMT) in Socorro, New Mexico, as its President. There were then 300 students at the Institute and well over 1000 by the time he left ten years later. Undergraduate students were employed in research jobs much as graduate students are at other universities. In addition to being President, he also continued his scientific research, publishing many papers with both graduate students and undergraduate students. The work with Chester McKee explained the origin of the light curve as powered by ^{56}Ni decay. NMIMT was a small institution, much the size of a reasonable physics department at other universities, so that doing science, being president, publishing and research was part of the job description. Stirling claimed it was easier to get money for scientific research proposals than trying to raise it as the usual college president does by gifts. It was also far easier to run an institution with the close collaboration of many students working with you on scientific projects then maintaining the august and often remote position of a college president.

An automated astronomy project was conceived in the fall of 1964 while traveling to an American Astronomical Society meeting in Tucson. In order to understand supernova, more of them needed to be found, and it seemed like an ideal project for a small institution in the southwest. NMIMT had a long tradition of both student employment and the use of surplus military equipment. A mountain laboratory already existed for atmospheric research started by Irving Langmuir and it was natural to add a telescope.

Stirling left NMIMT after ten years, and in December of 1974 joined Los Alamos Laboratory. It was a matter of significant achievement to Stirling that after starting as an electrician and engineer and spending a major fraction of his life as an experimental physicist, he finally made the transition to being hired in one of the most prestigious theoretical groups of the world, the Los Alamos Theoretical Division.

In this Division he is in charge of the Theoretical Astrophysics Group and has become a Senior Fellow in the Laboratory and is now a Member of the National Academy of Sciences. In the Theoretical Group he still works with his closest collaborator, Albert Petschek, who is now full—time at NMIMT but spends his summers and vacation time at Los Alamos with the Physics Division of the Laboratory. Stirling tends to have the ideas and Albert makes sure that the physics are correct. The computers of Los Alamos are a major resource for astrophysics. Analytical understanding is the bottom line.

SUMNER STARRFIELD

Sumner Starrfield was born in Los Angeles in 1940, and was raised in the San Fernando valley, just north of Los Angeles. As a high school student in the late 1950's, he was a volunteer observer for the "Moon Watch Program," which was run by the Smithsonian Astrophysical Observatory. When the first satellite went up (Sputnik), Sumner was out observing it. The next year, in 1958, Sumner went to the University of California (Berkeley) where he majored in mathematics. As an undergraduate at Berkeley, Sumner took a few astronomy courses—a nice application of math—and worked for a summer at the Hat Creek radio observatory in northern California, near Mt. Lassen.

Sumner, more than anything else, wanted to do independent research—application of mathematical ideas and models to interesting problems. Astronomy looked like a good area for this, and Sumner entered graduate school at the University of California at Los Angeles, majoring in Astronomy. Two events shaped Sumner's future direction. One was a talk on novae given by Jim Kaler, then a graduate student at UCLA. The other was the UCLA language requirement to translate a scientific work. Sumner translated Schatzman's early 1950 papers from French to English.

Schatzman, a German—Jew, escaped to France in the early days of World War II, and worked at the Pic du Midi Observatory high in the French Alps as a janitor. Immediately after the war, he applied the knowledge newly gained from nuclear weapons research to the novae problem.

Robert P. Kraft had just published a paper (in 1964) that suggested, on observational grounds, that many novae were members of binary systems—perhaps all novae, with one member of the binary being a white dwarf. Someone proposed that the novae explosion might be a result of hydrogen from the other star falling on the white dwarf, where it could accumulate and then go off like a gigantic H—bomb, but at the time Kraft did not think enough mass would transfer to do the job. Now here was a problem Sumner could sink his mathematical teeth into!

Sumner modeled a hydrogen layer on a white dwarf, writing the code for a computer simulation. He wanted to see if enough material could accumulate to do the job—a novae explosion—and found that indeed it would work, and thus Bob Kraft's objection did not hold. This work, which earned him a doctorate in 1969, launched Sumner as a theoretical astrophysicist, and he joined the faculty at Yale University, where he taught and did research from 1967—1971. At Yale, Sumner began his long association with Jim Truran—they still get together every summer, along with Stirling Colgate and Art Cox, at Los Alamos.

In 1972, Sumner accepted a position at Arizona State University. He was the first astronomer at ASU, and has built up a modest—sized astronomy section within the Physics Department over the last 15 years. Work on novae modeling led to the idea that perhaps if enough mass could be dumped on a white dwarf, it could be driven over the Chandrasekar limit, and the whole thing would explode as a Type I supernova. Ken'ichi Nomoto developed models that showed this was indeed possible.

CHAPTER 16

THE VANCOUVER MEETING

Formal published papers, clean theories that explain the observed facts, and textbooks that seem to cover everything, are science dressed up. There is another science, however, one of view graphs hastily put together in the heat of battle, talks that will never see print except as abstracts, open disagreements between scientists, old friends seeing each other again, rumors about who is getting what grant, and exotic places and fun—filled evenings. This is the world of science in the making, a world of people.

Nowhere is it easier to see science without its formal clothes on than at an annual meeting of a major scientific society. This is especially the case when something new and exciting is yet to be resolved. Such was the case with the 170th Meeting of the American Astronomical Society on June 14—18, 1987. Held jointly with the Canadian Astronomical Society at the University of British Columbia in Vancouver, this meeting featured a special session on Supernova 1987A that was organized by Andrea K. Dupree and Robert P. Kirshner.

The Vancouver meeting was not the first astronomical meeting to have a session devoted to Supernova 1987A. Several meetings had preceded it, some entirely devoted to the Supernova. The Vancouver meeting was, however, the largest gathering of astronomers since 1987A had exploded.

The American Astronomical Society (AAS) is the major organization of professional astronomers in the United States, Canada, and Mexico. The basic objective of the AAS is to promote the advancement of the science of astronomy (and related sciences, such as physics and mathematics).

The AAS meets these objectives in a number of ways, but one of the most important is the publication of journals such as the *Astrophysical Journal* and the *Astronomical Journal*. Another is to hold two general meetings each year, usually in January and in June. The winter meetings are very well attended. The meeting in Tucson in 1985 was attended by 2000 astronomers, one of the largest gatherings of astronomers ever. The summer meetings are not as heavily attended, but depending on the location and the mood of the times, they can draw a large number of astronomers. Such was the case of the 1987 summer meeting in Vancouver, British Columbia—held when interest in Supernova 1987A was at a high. Some 800 astronomers attended.

Figure 1. A tour to the Dominion Astrophysical Observatory on Vancouver Island treated astronomers with the sight of the second largest telescope in Canada.

Scientists who wish to present results have a choice (at an AAS meeting) of giving either an oral presentation or a poster paper. Science at such a meeting is very democratic. If you are a member, get your abstract in on time (the deadline is absolute!), and pay your $25 fee for abstract publication, you **will** be scheduled in an appropriate session.

Generally at the last minute, abstracts come flooding into the AAS Executive Office in Washington. Washington–based astronomers are drafted to help sort them into sessions and Pam Hawkins and the Executive Office staff help Peter Boyce make the many arrangements.

Oral presentations are allowed exactly five minutes (with a warning beep at four minutes). Whether the paper is of earth–shaking consequence with the audience on the edges of their seats, or is only of interest to one or two astronomers in the audience with the rest drifting off to sleep, the time is exactly the same—five minutes.

THE SPECIAL SUPERNOVA SESSION

Monday and Tuesday passed quickly with many papers and posters. On Monday evening there was a dinner cruise around the Vancouver harbor. On Wednesday morning, the special session on Supernova 1987A was jam–packed with astronomers. At the appointed time Andrea K. Dupree stood up and silence fell in the hall as astronomers settled down for the session.

Andrea stated straight off that Supernova 1987A was the "most important astrophysical event in the 1980s." "To some this supernova was the Holy Grail of astrophysics, while to others, it was merely a light bulb to illuminate the interstellar matter." Andrea pointed out that a new term had entered astronomy: the "supernova spouse" (the one left behind as the other pursued the supernova day and night). Andrea then introduced the first speaker, Robert F. Garrison, the Director of the University of Toronto Southern Observatory. Bob had hired Ian Shelton, the discoverer of the supernova.

The title of Bob's talk was "Supernova Shelton 1987A: The Discovery and the News Via Short–Wave Radio from Last Night." He pointed out that the discovery was not a lucky accident. Fate did not place a Canadian observer on a lonely mountaintop in the southern hemisphere; rather, an enterprising astronomy department had the foresight to know that it is important to have a window on the southern skies. Many people looked up at the LMC that night, but only Ian had enough background to know the significance of what he was seeing and the good judgment to know what to do with it.

Bob also pointed out how hard Ian had worked since he discovered the supernova. Some 112 nights of photometry of Supernova 1987A had been obtained at the University of Toronto Southern Observatory (UTSO). Of these, 73 nights had been obtained by Ian himself, in spite of a trip back to Toronto to celebrate the discovery.

The next speaker was Mark M. Phillips from Cerro Tololo Inter–American Observatory (CTIO), one of the observatories of the National Astronomy Optical Observatories (NOAO). The title of Mark's talk was "Optical and Infrared Observations of Supernova 1987A from Cerro Tololo Inter–American Observatory." Mark emphasized that most of the CTIO astronomers had been working on Supernova 1987A, and that he had been sent to speak for them.   Observations at CTIO by the resident staff included photometry in the *U, B, V, R,* and *I* bands (ultraviolet through the near—infrared on the hastily recommissioned 0.4—meter telescope), spectrophotometry, and infrared photometry.   Observations by visiting astronomers included CCD spectroscopy, speckle interferometry, and high—speed photometry.

Figure 2.   Andrea K. Dupree, Vice President of the American Astronomical Society and Director of the Solar and Stellar Physics Division at the Harvard—Smithsonian Center for Astrophysics, calls the special Supernova 1987A session to order and gives a few introductory remarks (Photo by Russell Genet).

Long—time CTIO astronomer, and former CTIO Director, Victor Blanco, examined many pre—outburst plates, and was able to establish that the suspected progenitor, Sanduleak $-69°$ 202, had not been a variable star—at least not noticeably so. He was also able to confirm that there had been three, not two stars, visible before the outburst. The CTIO astronomers were quite certain that the B3 supergiant, Sanduleak − 69° 202 was indeed the progenitor. They pointed out that there were many other B3 supergiants in the LMC, and suggested that they might get much more attention in the next several years.

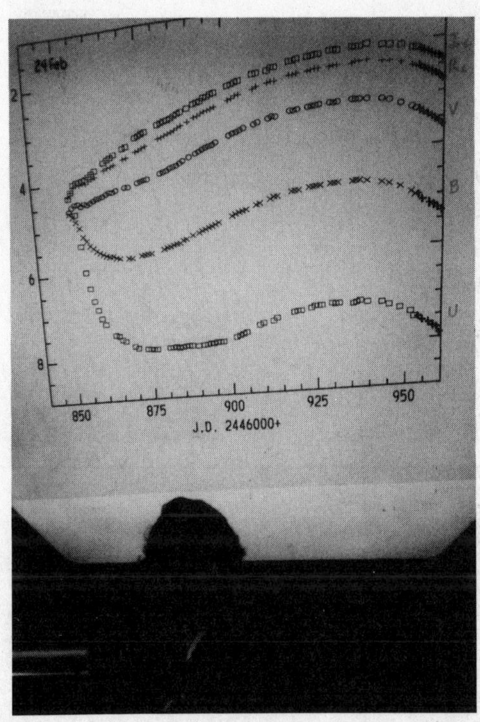

Figure 3.   Mark Phillips displays CTIO photometry of 1987A, clearly showing a recent drop in brightness in all the color bands.   The brightness increases upward, and time (days) increases to the right.

Andrea Dupree then introduced the next speaker as "the man with astronomy's most famous smile."   It was Robert P. Kirshner, from Harvard University and the Harvard–Smithsonian Center for Astrophysics, to speak on "Ultraviolet Observations of Supernova 1987A."   He noted that observing with the International Ultraviolet Explorer was great, "they had not had any problems with the weather!"

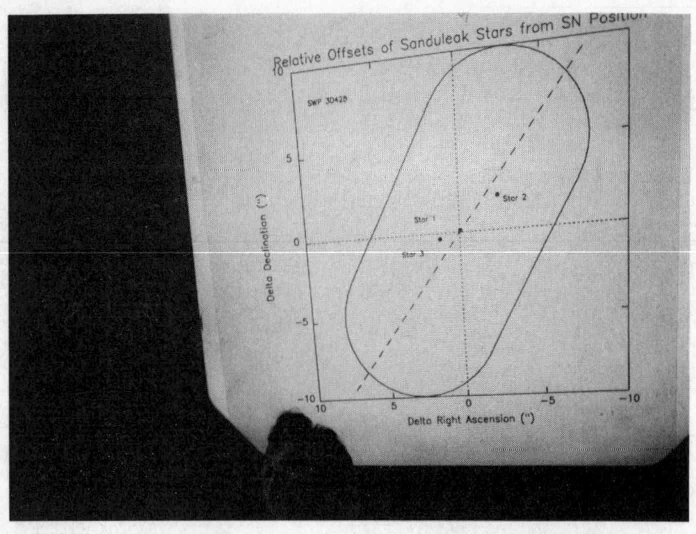

Figure 4 (previous page).   Top: Robert P. Kirshner explains that analysis of IUE data suggests that there were three stars originally, with two remaining.   Bottom: The first reference at the meeting to the "Mystery Spot" was a view graph Bob hastily flashed on the screen.

Bob pointed out that the ultraviolet radiation had fallen off very rapidly, and even though they had been able to make observations on the day of discovery, the ultraviolet had already peaked and was declining at the point of their first observations (20 hours UT, on February 24th). The ultraviolet spectrum looked initially very much like the spectrum of a Type I supernova, but Bob thought this was because the progenitor did not have much of a circumstellar envelope.

Figure 5.   Andrea Dupree straps the microphone on Daniel Sinclair, who spoke on the detection of neutrinos from Supernova 1987A (photo by Russell Genet).

Bob also pointed out that after the ultraviolet radiation from the supernova had declined, that the ultraviolet light from two stars, not one, was apparent ("in spite of what I may have said previously to the contrary" he joked).   He then flashed a bumper sticker from a Santa Cruz meeting—making the first fleeting reference to the "Mystery Spot."

Andrea Dupree then introduced the last speaker of the special morning session by pointing out the observations were moving from geosynchronous orbit (IUE) to subterranean realms (literally underground). Daniel Sinclair, from the University of Michigan, spoke on "Neutrinos from Supernova 1987A."

THE WEDNESDAY MORNING SUPERNOVA SESSION

After a half—hour for coffee and a chance to look at some of the posters, there were three parallel oral sessions. Astronomers had to choose which one they went to. Most choose Session 34: Supernova 1987A. Robert Wagoner, from Stanford University, called the session to order.

Richard N. Manchester, Commonwealth Scientific and Industrial Research Organization (CSIRO), Epping, New South Wales, Australia, was the first speaker, and he wasted no time getting to the podium. His topic was "Radio Emission from SN 1987A in the Large Magellanic Cloud." Richard reported that they made their first observations at radio wavelengths soon after Supernova 1987A was discovered, and were able to detect it, although it was not strong. It peaked about three days after discovery and then faded. This radio event may be but a small precursor to a major radio flare such as has been observed in other supernovae.

The next speaker was George Sonneborn, the International Ultraviolet Explorer (IUE) Resident Astronomer. His topic was "Spatially—Resolved Far Ultraviolet Spectroscopy of the Field Around Supernova 1987A." George gave details on what Robert Kirshner had mentioned in his invited talk, namely that they had been able to use the IUE to determine the spatial orientation of the two remaining stars with respect to the supernova. What they found was consistent with what had been found at European Southern Observatory and Cerro Tololo Inter— American Observatory; that Sanduleak −69° 202 was the progenitor to the supernova, and that two other stars there before the explosion were still there, while Sanduleak −69° 202 was not.

Next, Andrea K. Dupree (Harvard—Smithsonian Center for Astrophysics), and then Fred C. Bruhweiler (Catholic University of America) reported on the interstellar medium towards Supernova 1987A. Andrea described a number of different absorption lines along the line— of—site to the supernova that originated in the LMC and in our own galaxy. She and her colleagues discovered an excess of highly ionized atoms towards the supernova. Fred concluded that the Supernova 1987A explosion occurred in or near a hole or "cavity" caused by a region full of hot, ionizing O—type stars. Perhaps the intense radiation from these stars had essentially cleaned out the area.

Figure 6.   Top left: Robert Wagoner leads the session.   Top right: Richard Manchester from Australia, reports on prompt radio emissions.   Bottom left: Edward Chupp reports no prompt gamma rays detected by the Solar Maximum Mission spacecraft.   Bottom right: Kenneth Brecher predicts a pulsar.

Edward L. Chupp, from the University of New Hampshire, reported on the Search for Prompt Gamma—Ray emission from Supernova 1987A. The Solar Maximum Mission—the spacecraft repaired by astronauts while in orbit—had a gamma ray detector aboard, and it was operating at the time a gamma ray burst would have arrived from the supernova. The Solar Maximum Mission spacecraft was not, of course, pointing the gamma ray detector at the supernova at the time, but a gamma ray burst would have been detected (the supernova had not been discovered yet), as the gamma rays would have penetrated the detector from the side. No gamma ray burst was detected.

S. M. Matz, from the Naval Research Lab, reported on A Search for Gamma Ray Lines from Supernova 1987A in the LMC. Matz, like Chupp, had used data from the Solar Maximum Mission spacecraft, although he was not looking for a gamma ray burst, but was instead was looking for a steadier emission. No specific lines were detected yet, but he pointed out that the expanding supernova shell was expected to be opaque to gamma rays for some time. Typical supernova shells were opaque for about a year. However, Supernova 1987A was not typical, and because there was presumably less circumstellar matter at the time of the explosion (being a more compact blue instead of an extended red giant), the gamma rays might shine through much sooner—perhaps in just a couple of months.

Kenneth Brecher, from Boston University, spoke on "Gamma Rays From Supernova 1987A." A theorist, Kenneth predicted that gamma rays would be seen from Supernova 1987A as early as a few more months and as late as two years. He suggested that gamma rays might be seen from decaying atoms, synthesized in the explosion, from protons accelerated in the supernova remnant, from positron annihilation, and from the resulting neutron star/pulsar. Kenneth predicted that we would see a pulsar in 2 to 6 months flashing with a period of 2 to 6 milliseconds if we were lucky enough to have the beam pointing in our direction.

The next speaker, Eli Dwek, from NASA's Goddard Space Flight Center, had good news/bad news. The good news was that IR observations would be able to determine many important characteristics of the area into which the supernova was expanding. The bad news was that when the ejecta from the supernova condensed into dust particles, the supernova might become opaque in the ultraviolet and optical (but not the infrared) for as long as three years—blocking out all the exciting observations everyone else was planning on making. The opaque period could start as soon as 130 days after the explosion. That dust forms from supernovae is a certainty—supernovae are the main source of dust. Eli showed a view—graph with a robed and bearded prophet out of biblical times carrying a sign that said "Ultraviolet Astronomers Repent"—alluding to the fast decline of the ultraviolet and the rise of the infrared.

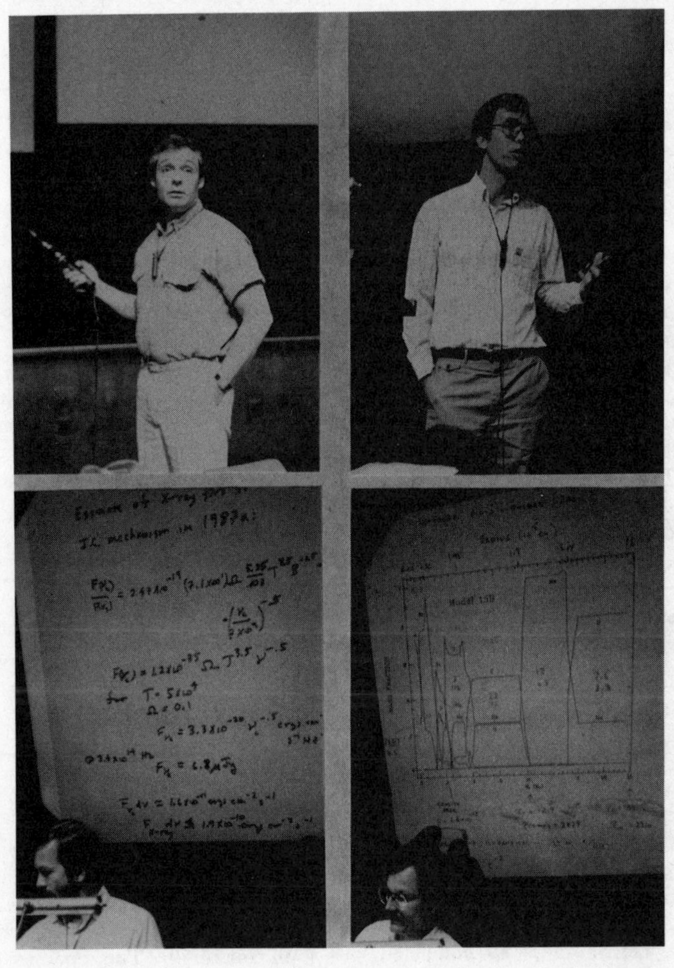

Figure 7 (previous page).   Top left: Eli Dwek predicts that dust will soon obstruct most observations of Supernova 1987A.   Top right and bottom left: X—ray astronomers Francis Marshall and James Beall discuss what might be expected in the X—ray region.   Beall proved his point with a few equations.   Bottom right: Richard McCray confidently predicts that X—rays will be visible this fall and that the ultraviolet will return—at least that is what Stan Woosley's Model 15B, shown on the screen, suggested.

The next two speakers, Francis Marshall (NASA's Goddard Space Flight Center), and James Howard Beall (Naval Research Laboratory) addressed the likely X—ray situation with Supernova 1987A.   Frank summed up the current situation nicely when he said "I have no data, of course, being an X—ray astronomer!" (The United States has not had an X—ray telescope in space for years).   Frank explained that when the shuttle got going there would be some X—ray observations, as well as some from small rocket flights.   Jim explained what might be expected in terms of X—rays from Supernova 1987A.   Jim also had the most amusing view—graph of the session.   It showed the earth as a globe, with astronomers rushing from the northern to southern hemisphere, and the concentrated mass of astronomers in the southern hemisphere upsetting the balance of the earth.

The final speaker in the session was Richard McCray, from the University of Colorado.   The provocative title of his talk was "Inside Supernova 1987A."   He used a model to predict that Supernova 1987A will become bright in the X—ray portion of the spectrum this fall—bright enough for the Japanese X—ray telescope, "Ginga", now in orbit, to see with ease.   Dick also predicted that there would be a renaissance in the ultraviolet this fall for the IUE observers.

THE PRESS CONFERENCE

It was one of the biggest and most exciting press conferences ever run by the American Astronomical Society.   The room was full of prestigious science reporters from magazines, newspapers, and television.

Stephen P. Maran, the American Astronomical Society's Press Officer, ran an efficient, well organized press conference.   He had selected a dozen astronomers as panelists, and between them they were expert on every aspect of Supernova 1987A.   Stephen Maran made a few opening remarks and introduced the panelists.

Bob Garrison, the first of the two planned speakers, summarized the discovery of the supernova, and its significance.   He emphasized that when Shelton saw the supernova, he knew what it meant and knew what to do with it.   It was the brightest supernova in 383 years—the last one was observed before the telescope was invented.   It was particularly significant, Bob thought, that the precursor star, Sanduleak −69° 202 had been observed *before* the supernova explosion.   Bob, a man most appreciative of the beauty in astronomy, reported that visually the supernova was now a most gorgeous red against the silver—grey of the Large Magellanic Cloud.

Figure 8 (previous page).   The panelists from top, left to right.   Ken Brecher, Ed Chupp, Andrea Dupree, Bob Garrison, Bob Gehrz, and Bob Kirshner.   Bottom, left to right.   Dick McCray, Richard Manchester, Mark Phillips, Daniel Sinclair, and Stan Woosley.

Peter Nisenson, from the Harvard—Smithsonian Center for Astrophysics (CfA), was the second and last planned speaker.   Peter summarized how he had taken the speckle camera he had developed to the 4.0—meter telescope at CTIO and made observations on March 25th (38 days after the explosion was observed).   He was trying to measure the diameter of the rapidly expanding supernova to determine the position of the supernova with respect to other nearby stars.

The speckle camera results can not be seen immediately, but must be processed by a computer at the CfA.   When this was done, Peter and his co—workers were shocked to see a second object next to the supernova, only about two light weeks away.   It was measured on two separate nights and seen on both nights.   It could not be equipment problems, because the Anglo—Australian Telescope had seen it also, at the same position.   The the "mystery spot", or "speckled thing", or whatever it was called, had to be real.

Steve Maran, the AAS Press Officer, opened the floor to questions.

Linda Monroe, from the *Oregonian*, asked Peter Nisenson, "What is it?"

Peter said that it was "unique, hard to say what it was, every explanation proposed so far had serious problems."

The *Science News* reporter, Detrick Thompson wanted to know what a theorist thought, and this was directed to Stan Woosley.   Stan pointed out that it was not there before the supernova (too bright for that).   It was *not* a second supernova (he would have to quit astronomy if it was a second supernova!)   It had to be receiving its energy from the supernova, yet it was very bright for its size to be simply reflecting the light from the supernova.

Richard Manchester, the Australian radio astronomer, suggested the initial burst of neutrinos had set off a nearby explosion.   Stan Woosley immediately responded with a brief "That's impossible!"

Kathy Sayer, from the *Washington Post*, wanted to know how the mystery spot could be so bright.

Peter Nisenson pointed out that even if the mystery spot was a perfect reflector pointed at the earth, this could not explain how bright the mystery spot was unless the energy from the supernova were somehow *beamed* at it.

Ken Brecher quipped "Beam me up Scottie!" and suggested that it was an X—ray or Gamma ray beam, or a fast particle beam moving at half the speed of light from the newly created pulsar.

Richard Manchester pointed out that if it were a beam from a pulsar, it was extraordinarily bright, 10,000 times as bright as the pulsar in the Crab Nebula.   Admitting that he had never seen such a young pulsar, Richard said it was "still hard to stomach" such a bright pulsar beam.

Figure 9.   Top: Peter Nisenson, an astronomer from the Harvard–Smithsonian Center for Astrophysics that led the team that discovered the mystery spot.   Bottom: Ken Brecher, the theorist from Boston University.

Stan Woosley said he predicted a neutron star (which could be a pulsar) and that it would have a mass of 1.35 solar masses, give or take 0.2 of a solar mass.

If a pulsar was flashing in our direction, Richard Manchester pointed out that they were already looking for it at both the optical and radio wavelengths in Australia.

Andrea Dupree pointed out that the supernova exploded in a very complex region of space. "There is lots of ionized material there," she said.

Andrea pointed out that it was a sad commentary on the state of the United States space astronomy program that the only telescopes available to observe the supernova were "space antiques." The IUE was launched in 1978, while the Solar Max was launched in 1980, she pointed out. Then she smiled and said, "The good news is that they were able to fix the Solar Max, and that both satellites have lasted so long.

Bob Kirshner commented that the long, slow rise in brightness, caused either by energy from the pulsar or radioactive decay, was now over, and the supernova was now declining and would be harder to observe. It will take longer exposures, and it will become harder to take telescope time away from other previously planned projects. "Of course, if as predicted by Robert Gehrz, the dust blocks out everything, the we won't have to worry about observing at all!" Bob suggested.

Richard McCray, the high—energy theorist, admitted, "It doesn't make sense. Nature is more inventive than we astronomers are!" Richard pointed out that the expanded supernova was moving at one tenth to one twentieth the speed of light. This means that in a few months the leading edge is going to slam into the mystery object. "If there is an object there, we should see quite a blaze!" he said.

Mark Phillips, the CTIO spectroscopist, pointed out that while the mystery object could be an emission line object (such as a cloud of gas), it seemed unlikely as the emission lines would already have been detected.

Peter Nisenson noted another difficulty with the idea it was a simple reflection off a gas cloud. The mystery object was much redder than the supernova. As there was a time lag of a couple of weeks between the supernova and the mystery object (which was a couple of light weeks away), and as the supernova was getting redder, one would expect the mystery object to *lag* the supernova in time, and would expect it to be *bluer*, not redder.

Linda Monroe, from the *Oregonian*, asked Robert Gehrz "Would dust really cut off the ultraviolet and optical radiation?"

Robert said that in other supernovae, dust had often greatly reduced the intensity—sometimes by a factor of 100 (a very large drop of five stellar magnitudes). Robert said that "If we don't see dust we should be surprised. The onset of the dust could be as early as August. There may be an abrupt drop in light that could last from six months to years, a nuclear winter."

Karl Hodge, from the *Arizona Republic*, asked Richard McCray to expand on what would happen when the supernova ran into the mystery object.

Richard answered that most of the energy would be in the form of X—rays and gamma rays. "The Japanese satellite should detect it first," he said.

"What," asked one of the reporters, "did it mean if neutrinos had mass?"

Ken Brecher replied that the universe was filled with neutrinos. If, as suggested by Hong—Yee Chiu, an astronomer from the NASA Goddard Space Flight Center, neutrinos had a mass of 3.5 electron volts, this would be a large mass in the universe, but still not large enough to "close" the universe—to keep it from expanding. For this to happen, the neutrinos would need to have a mass of about 20 electron volts.

Stan Woosley expressed doubt about probable masses for neutrinos. Upper limits to the mass were all right, but he was not convinced that evidence really suggested a value different than zero.

Hong—Yee Chiu said he was not there to defend his assertion that the most probable value of the mass for the neutrino was not zero, but was 3.5 electron volts, though he gave a rigorous defense the following day.

On of the astronomers suggested that people have this fantasy that the can close the universe with neutrinos with mass. We should rename them "Fantinos!" Time was up and Steve Maran adjourned this exciting press conference.

CHAPTER 17

THE NEXT SUPERNOVA

When we ask about the next supernova, now that Supernova 1987A is history or at least part of the on—going present, there are several ways we can formulate the question. (1) When and where will the next supernova which is bright enough to see with the naked eye explode? (2) Which is the nearest star to Earth which we think may be slated for a supernova explosion some time in the future? (3) What are the prospects for a really nearby (hence really bright) supernova, the effects of which could be catastrophic to inhabitants on the Earth's surface, or perhaps alter the course of evolution? Let us consider these three questions in turn.

A supernova explosion bright enough to be visible to the naked eye would have to occur in our own galaxy, in the Large or Small Magellanic Cloud, or in one of the nearby members of our Local Group of galaxies. A supernova in our own galaxy could be extremely bright (brighter than the full moon) if it occurred quite nearby, or it could be extremely faint or even invisible if it happened to occur in a distant part of our galaxy behind the large clouds of interstellar dust which obscure starlight. A supernova in either of the Magellanic Clouds should reach magnitude -1 or 0 depending on whether it was of Type I or II, respectively. This was the maximum brightness predicted for Supernova 1987A, which in fact puzzled and disappointed astronomers by reaching only magnitude $+2.9$. A supernova in M31, the Andromeda Nebula, would reach 4th or 5th magnitude depending on whether it was of type I or II, respectively. Such was the case of the last one known to occur in that galaxy, which reached magnitude $+5.8$ in August of 1885.

Statistically, supernovae should occur more frequently in larger galaxies because they contain more stars. The overall rate, lumping Types I and II together, is about one explosion every 20 years for large galaxies such as ours and M31, one every 100 years for medium—sized galaxies like M33, one every 250 years for the Large Magellanic Cloud, and one every 1000 years for the Small Magellanic Cloud. Thus, if you would appreciate a "hot tip," we would advise placing bets split evenly between our galaxy and M31.   With only 100 years since the last supernova in M31 but over 380 years since the last supernova visible in our own galaxy, our galaxy would seem to be the better bet, except for the fact that the interstellar dust could make for a repeat performance of the 1667 supernova, which was never seen.

When asking what is the next star which astronomers think might undergo a supernova explosion, most people think of Betelgeuse, otherwise known as Alpha Orionis, the brightest star in the constellation of Orion (the Hunter).   Betelgeuse is a massive, cool, very large, highly luminous supergiant.   Its large mass predestines it for a supernova explosion of Type II.   Moreover, its present configuration as a large, cool supergiant tells astronomers it is near the end of its total lifespan of 10 million years.   Thus we can say that sometime within the next one million years, give or take a few years, Betelgeuse will explode.

Its distance from Earth is about 300 light years.   Although it is at present moving slowly away from the Earth   (at a rate of about 20 kilometers per second), its distance should not increase more than about 10 or 20 percent during the next million years we are waiting.

At a distance of 300 light years, a supernova of Type II would reach a peak apparent brightness of about −13 magnitudes.   This is considerably brighter than any of the historical supernovae listed in Chapter 3, even the 1006 A.D. supernova in Lupus, which reached magnitude −9.5.   For comparison, note that −13 is the apparent brightness of the full moon.   Any attempt to describe the drama unfolding once the Betelgeuse supernova occurs would be a series of understatements.

The attention focused on Betelgeuse might, however, be misplaced. One of the major surprises which Supernova 1987A threw at us was that its progenitor was not the expected cool, very large supergiant but, it was a hot, only moderately large supergiant.   If a massive star can explode while still a hot supergiant, and Sanduleak −69° 202 certainly did, then it would be prudent to keep the corner of one eye on other nearby hot supergiants as likely candidates for the next nearby supernova explosion. One such star is Rigel, otherwise known as Beta Orionis, the second brightest star in the constellation of Orion.   It is certainly massive enough that it must become a supernova sometime and, at a distance of 900 light years, it is not much farther away than Betelgeuse.

In the discussion above we have considered only supernova of Type II, but in fact supernovae of both types, I and II, occur in roughly equal number in most galaxies.   Keeping an eye on precursors of Type I explosions is, however, much more difficult.   That is because, according to the best current theory, they are white dwarf stars very close to the Chandrasekhar limit and about to be pushed over the limit as a result of mass transfer from a nearby companion star in orbit around it. Supergiants are extremely luminous (the most luminous known stars in a galaxy), standing out prominently as bright beacons—advertising their presence and exact location.   White dwarfs, on the other hand, are among the least luminous stars in a galaxy.   Most white dwarfs are many millions of times fainter than supergiant precursors of Type II supernova. Moreover, it is not an easy matter to know which of the thousands of millions of white dwarfs in a galaxy are very close to the Chandrasekhar limit and on the verge of being pushed beyond it. So, don't be surprised if the next star to explode and attain naked eye brightness is not one of the supergiants you were watching (or betting on).

Most massive supergiants, the candidate stars ripe for supernova explosions of Type II, reside in the spiral arms of our galaxy. Every spiral galaxy, ours included, has basically two major spiral arms structures, if we ignore minor spurs, protrusions, and irregularities. We (the Earth, our Sun, and the rest of our solar system) orbit around the center of our galaxy once every 200 million years or so. During each full orbit we enter into, pass through, and emerge from each of the two major spiral arms. Thus, every 100 million years or so, as we make the passage which lasts about 10 million years, we come into proximity with a relatively large number of those ticking time bombs. There is a good chance that, during each 10–million–year–long passage, a supernova will explode as close as 30 light years. During the five thousand–million year age of our solar system, there should have been around 50 supernova explosions which occurred within about 30 light years of the Earth. Of these, some would have been farther, some closer than 30 light years away. A few could have been as close as four light years, the distance separating us today from our nearest neighboring star—the Proxima/Alpha Centauri triple system. Let us explore the consequences of a supernova which might, sometime in the (relatively distant) future, occur at that proximity.

If a star four light years away terminated its life in a supernova explosion, we would have no inkling at all of the catastrophe until four years later. Nothing can travel the distance of four light years in less than four years time, and only photons and neutrinos can do it in four years.

The first message would arrive in the form of a brief but incredibly intense shower of neutrinos, picked up without ambiguity by any and all neutrino detectors in operation at the time. The onset of the shower would be as sudden and its duration as brief (10 seconds or so) as was the shower detected at 7:35 A.M. Universal Time on 23 February 1987 from the supernova in the Large Magellanic Cloud. The intensity, however, would be two thousand–million times greater. Despite the phenomenal intensity, however, the neutrino shower would do virtually no damage to living organisms, because of the remarkable ability of neutrinos to pass through the atoms in solid objects (such as human bodies) without interaction.

The star, or what had been the star, would begin increasing in brightness almost immediately thereafter but would take a full 24 hours to climb to near its maximum brightness. The climb itself would be breath–taking indeed: doubling or tripling in brightness every hour. The gravity of the situation would be seen once it reached its maximum brightness that next day. A supernova of Type II, at that distance, would reach an apparent magnitude of -22.6.

Since our own Sun has an apparent brightness of −26.7, this nearby supernova would be about one fortieth as bright as the Sun, but it would be truly spectacular when seen in the night sky. The increased flux of radiant energy falling on the earth would heat up the entire surface of the globe by only a small amount (only about two degrees Centigrade) during the time of maximum brightness. For thousands of millions of years our Sun's remarkably steady flux of radiant energy has kept the Earth's surface at an average temperature (the average over night and day, winter and summer, north and south) of about 20 degrees Centigrade, ideally suited for most of terrestrial life. It would be quite an exercise in climatology and ecology to determine precisely what would follow next, but it would not likely be disastrous. In any case, the supernova would be at its brightest for only a couple of months. After one or two years it would still be as bright as the full Moon, but its ability to cause further damage by overheating would have ended.

The effects of cosmic rays would be another matter. During and after the melee of the supernova implosion/explosion catastrophe which took place four light years away, copious amounts of cosmic rays were spawned. We are already familiar with cosmic rays, which are responsible for about 1/3 of the natural background radiation present at all times on the Earth's surface, the other 2/3 emanating from various radioactive materials in the Earth's solid crust. Cosmic ray physicists and stellar astrophysicists have already determined that virtually all of the cosmic rays, which emanate from all parts of the sky, had their origin in the multitude of supernova explosions which have occurred in various times past and are now finding their way through interstellar space to escape from our galaxy into intergalactic space. What are cosmic rays? They are actually particles, not rays, those particles being bare nuclei of atoms which have been stripped of the electrons which normally surround an atom in its neutral state, and the free electrons thus removed. The most common cosmic ray particles are the free electrons (historically called beta rays), the bare nuclei of hydrogen atoms (which are simply protons), and the bare nuclei of helium atoms (historically called alpha rays). For completeness we mention that gamma rays, historically included as one component of cosmic rays, are indeed rays: they are simply photons with extremely short wavelengths and high energies. Cosmic ray particles spawned in a supernova event are imparted with tremendous energies and consequently extremely high speeds, some approaching the speed of light. No particle of matter which has mass can ever travel with exactly the speed of light; photons and neutrinos, which have no mass, can and must and do travel with exactly the speed of light. The fastest of these newly born cosmic particles, those traveling at, say, 90% the speed of light, will cover the four−light−year distance in four years and a few months. The somewhat slower cosmic particles will continue to follow in greater numbers and reach the Earth in subsequent months, years, and decades. The result will be increased levels of radioactivity on the Earth's surface, most of it now coming down from the sky rather than up from the ground. The intensity will be so great as to spell virtually immediate death for most life forms. Radiation hazard is usually measured in terms of exposure (intensity multiplied by duration), the same formula used in photography. So, the hideous radiation intensity combined with its prolonged duration (decades or more) will combine to produce total exposures which would leave little hope for survival of individual organisms or even entire species. One can speculate whether certain organisms in the ocean depths or far underground might hang on, but there is no doubt that life on Earth and its subsequent evolution would suffer a monstrous set—back which would require at least a thousand— million years to reverse.

The less energetic and less rapidly ejected particles, those making up the expanding envelope of the supernova remnant, would reach Earth later. The reader will recall that spectroscopic measurements of Supernova 1987A showed that the envelope began expanding at speeds of 30,000 kilometers per second, which corresponds to 10% the speed of light. Corresponding particles from a supernova four light—years away if not slowed down in their interstellar flight, would reach Earth several centuries later. Subsequent envelope expansion, after a month or so, settled down to a leisurely 3000 kilometer per second rate, meaning that the bulk of the material would take several millennia to reach Earth. During those centuries and millennia, as the supernova remnant sweeps past the solar system, the entire sky will glow much brighter than the Milky Way. This will be a spectacularly beautiful and mercifully harmless display of pyrotechnics.

Despite the success of the basic idea of gradual evolution of life forms on Earth, namely, Darwin's Theory of Evolution and its later refinements and applications, there has been a gradual accumulation of evidence pointing to the eventual realization that several times, during the history of life on Earth, evolution was punctuated by "discontinuities". At those points in time numerous entire species of life suddenly disappeared and were replaced by brand—new species. This startling fact was not immediately obvious all along because the geological record, strata of fossil remains laid down in successive sedimentary layers gradually over the millions of years, is not capable of pin—pointing changes precisely in time even if they were in fact global, catastrophic, and instantaneous. The one global extinction referred to most often, although there have in fact been several, was the one which occurred 65 million years ago and caused the great dinosaurs (as well as about half of the other species then living) to disappear forever.

Starting with a 1957 paper by the Russian astrophysicist Iosef S. Shklovsky, and continuing up through the early 1970's, the idea was put forth that nearby supernova explosions were the culprit. It was argued that the radiation damage caused death to large numbers of individual organisms, induced fatal mutations in offspring, and accelerated evolution as a result of more frequent non—fatal mutations. Most of the supernovae in question, as that theory contended, were not as close as the four—light—year example described above, but would have caused increases in terrestrial radioactivity significant enough to explain the observed phenomenon of recurrent global extinctions.

NAME INDEX

SUBJECT INDEX

ABOUT THE AUTHORS

Russell M. Genet

Born in 1940, Russell Genet was raised in the small California town of Yucaipa in the San Bernardino mountains. He received his B.S. in Electrical Engineering from the University of Oklahoma, and an M.S. from the Air Force Institute of Technology. He has worked for the Air Force for over 30 years, primarily in research and development where he has been responsible for a new missile guidance system, space experiments, and new analytic techniques in operations research. For the past decade, he has conducted his research at the Air Force Human Resources Laboratory where he is currently working on low—cost simulators for training fighter pilots for combat. He is an instrument—rated pilot and flight instructor. Russell and his wife Ann have five children and make their home in Mesa, Arizona.

In 1979 he founded the Fairborn Observatory which was devoted to making photoelectric observations of variable stars. In 1980, with Dr. Douglas S. Hall, he founded the International Amateur—Professional Photoelectric Photometry (IAPPP) Association, and its quarterly journal, the *IAPPP Communications*. The IAPPP has grown to some 750 members from 45 countries. With Louis J. Boyd, Genet has developed automatic photoelectric telescopes (APTs), and the Fairborn Observatory in conjunction with the Smithsonian Institution now operates a number of APTs on Mt. Hopkins in southern Arizona. In the area of astronomy and use of microcomputers in astronomy, he is author or editor of a dozen books. Russell Genet is a recipient of the Astronomical Society of the Pacific's *Amateur Achievement Award* and the Astronomical League's *Leslie C. Peltier Award.* He is a member of the AAS, AAVSO, ASP, and IAPPP.

Donald S. Hayes

Donald S. Hayes holds a B. A. in Astronomy from Pomona College, and M. A. and PhD degrees in Astronomy from the University of California, Los Angeles. He has been on the faculty of Rensselaer Polytechnic Institute and Arizona State University, and on the staff of the Kitt Peak National Observatory. A member of the International Astronomical Union, the American Astronomical Society and the International Amateur—Professional Photoelectric Photometry (IAPPP), he is an author of many papers in photometry and spectrophotometry. He has been chief editor or a co—editor of the proceedings of six professional symposia, including *New Generation Small Telescopes*, and is a co—author of:

Supernova 1987A: Astronomy's Explosive Enigma. He is with Fairborn Observatory and the Institute for Space Observations, and his research interests include the automation of photometric telescopes and several topics in stellar photometry. Currently, Donald and his wife Julie make their home in Scottsdale Arizona.

Douglas S. Hall

Born in Lexington, Kentucky in 1940, Dr. Hall lived there until he went to Swarthmore College and got his B. A. in Chemistry in 1962. Realizing a change of interest, he went on to Indiana University to earn his M. A. in 1964 and Ph.D. in 1967, both in Astronomy. Ever since then he has been at Vanderbilt University, where he is now Professor of Astronomy and Director of the Dyer Observatory.

Dr. Hall is very productive in research, having published almost 300 papers, given many invited lectures at other institutions, and participated in a number of colloquia and symposia in the United States and abroad. His research has involved photoelectric photometry of peculiar binary and variable stars, coordinating a world—wide network of amateur astronomers doing similar work with small telescopes in their back yards, and pioneering the use of computer controlled automatic telescopes. He is best known for his photometry of BM Orionis, an eclipsing in the Trapezium in the Orion Nebula, which showed one star to be highly flattened by rapid rotation, for proposing that huge starspots explain the many peculiarities of the puzzling RS Canum Venaticorum—type binaries, and for co—founding the I.A.P.P.P., and for editing its journal ever since then.

David R. Genet

David R. Genet received a B. S. in Industrial Engineering from Southern Illinois University in 1985. Since then he has been employed in the Simulations Department of a major aerospace firm. He was a co—editor of *New Generation Small Telescopes* and chief editor of the *Photoelectric Photometry Handbook* as well as co—author of *Supernova 1987A: Astronomy's Explosive Enigma.* He is a member of the IAPPP and a publisher associate member of the AAS. In addition to writing and publishing books in the area of astronomy, he enjoys flying airplanes, firing Scandanavian pottery on his back porch during the summer months, and complaining about the heat with Debbie, the chief pilot for Dave's Lear Jet. David, his Siberian Huskie Sasha, and his cats, Cheapet and Punkinhead, make their home in Gilbert, Arizona.